12.50

The Stars Belong to Everyone

Helen Sawyer Hogg

DOUBLEDAY & COMPANY, INC.
New York

The Stars Belong To Everyone

How to Enjoy Astronomy

1976
DOUBLEDAY CANADA LIMITED
Toronto

ISBN: 0-385-12302-7

Library of Congress Catalog Card Number: 76-3370

FIRST EDITION

Printed and bound in Canada by T.H. Best
Printing Company Limited

Design by Maher & Garbutt Ltd.

Credits

Plate 1. Taken by Dr. Peter M. Millman, December 9, 1968.
2. Courtesy of The American Museum of Natural History.
3. Original photos supplied by the Surveys and Mapping Branch, Department of Energy, Mines and Resources, Government of Canada.
4. Taken by E.E. Barnard. Yerkes Observatory photograph.
5. Courtesy of Hale Observatories.
7. Courtesy of Hale Observatories.
8. Taken by Dr. and Mrs. Donald A. MacRae.
9. Taken by H. Gordon Solberg, Jr.
10. Courtesy of Hale Observatories.
11. Courtesy of Jet Propulsion Laboratory, NASA.
12. Left: copyright by the National Geographic Society, Palomar Sky Survey. Reproduced by permission of the Hale Observatories. Right: Taken by Alan Irwin.
13. Courtesy of Hale Observatories.
14. Courtesy of Dominion Astrophysical Observatory.
15. Courtesy of Dominion Astrophysical Observatory.
17. Courtesy of David Dunlap Observatory.
18. Courtesy of Dominion Astrophysical Observatory.
21. Courtesy of Hale Observatories.
22. Taken by Dr. Nolan Walborn.
23. Courtesy of Harvard College Observatory.
24. Courtesy of Harvard College Observatory.

*To those members of my family
in three generations
and my teachers
who have inspired and encouraged
my love of the stars*

Foreword

During the thousands of years that man has walked upright on this earth, he has looked up with admiration at the beauties of the heavens around him. For less than four centuries of this vast time he has had the advantage of a telescope to help him interpret the lights in the sky. Although the telescope has added enormously to our understanding of the heavens, we should not make the mistake of thinking that we must always have one at hand to enjoy the sky. On the contrary, I hope that this book will show that anyone can follow the beautiful and interesting events in the sky without a telescope. For the stars belong to everyone.

The fact that there is much to enjoy in astronomy without a telescope — just with your own two eyes — was brought home to me forcibly some years ago. On leave of absence from my regular position, I traveled to Washington for a year's astronomical work of an administrative type. The first day in my new post I was interviewed by a reporter who asked brightly, "Have you brought your telescope to Washington with you?" I had to confess I had not. "My telescope" is a little job weighing 40 tons, with a revolving shelter weighing 80 tons. But after the reporter had left, I realized that I *had* brought to Washington my capacity to enjoy the heavens. The beautiful phenomena attending moonrise and moonset and sunrise and sunset, the annual showers of shooting stars, Venus as evening star, the old moon in the new moon's arms — all these and many more were just as enjoyable and just as attainable for me as they would be if a giant telescope were standing at my side.

Many people tend to postpone their enjoyment of the stars because they are constantly with us, but the iridescence of the twilight bow or the orange harvest moon rising slowly over a smoky fall landscape are celestial real-life scenes to look forward to from night to night or year to year. Very little time is required to see and enjoy the beauties of the sky; once you come to know them, they never lose their appeal.

This book is an outgrowth of 30 years of lecturing, 25 years of writing a weekly column for a major newspaper, 20 years of poring over old books. It answers many of the questions that I am asked over and over. It is not intended as a textbook on astronomy — there are plenty of good ones available. Though it does describe the principal bright stars and constellations, it is not intended as a comprehensive study of constellations — there are many good star guides and atlases.

The book attempts to give explanations for celestial happenings. If some parts do not appeal to you, use Sir Winston Churchill's solution and skip them. While the chapters follow in a logical progression, if you prefer you can select those which most interest you for first reading.

The magnificence of nature is all around and above us. I hope this book will add to your enjoyment of it.

Helen Sawyer Hogg
October, 1975

Acknowledgements

I wish to express appreciation to various institutions and individuals who have aided my efforts for this book. To the University of Toronto which, during my forty years on the staff, has encouraged its professors to work with the general public; in particular to the present and past Directors of the David Dunlap Observatory, Professor Donald A. MacRae, Professor John F. Heard, and my husband, the late Professor Frank S. Hogg; and the librarians, and photographer Warren Magill for much help with the illustrations; to the *Toronto Star,* a great city newspaper which for 25 years has given space and support to my weekly column "The Stars"; to A. Milton Runyon for years of faith that this book would fill a small gap in the many books available on astronomy; to various members of the Royal Astronomical Society of Canada, the Society editor, Dr. Ian Halliday, and the editor of the *Observer's Handbook,* Dr. John R. Percy, for help with material; to Dr. R.L. Duncombe, U.S. Naval Observatory, for data on planets and eclipses; to Dr. L.W. Fredrick and D. Van Nostrand for permission to use the star maps from the excellent textbook Baker and Fredrick, *Astronomy;* to R. Newton Mayall and G.P. Putnam's Sons, Coward, McCann & Geoghegan, Inc. for use of the copyrighted Perpetual Star Time-Piece, Compass, and Calendar; to Dr. K.O. Wright for material from the Dominion Astrophysical Observatory; to Dr. Allan Sandage for Hale Observatories material; to Dr. Peter M. Millman for supplying many bits of information; to Dr. C.C. Costain for the Time Zone Map; to Dr. Walter Tovell for the copy of the Leonid Shower; to Betty Jane Corson of Doubleday Canada for valuable suggestions and patience in editing; to Frank Stevens for preparation of the drawings; to the McLaughlin Planetarium for useful material; to *Sky and Telescope* and *Science News* for their excellent coverage of new astronomical developments; to my sons Dr. David E. Hogg for careful perusal of the manuscript and James S. Hogg for comments, and my daughter, Mrs. Sally L. MacDonald, for all the typing and extensive checking; and to individuals and institutions cited in the figure and photograph captions.

H.S.H.

Contents

The Stars Belong to Everyone

★

★

★

ONE

The Bag of Tricks in the Earth's Atmosphere

Things are seldom what they seem. Skim milk masquerades as cream. High lows pass as patent leathers. Jackdaws strut in peacock's feathers. Very true! So they do!

W.S. GILBERT, H.M.S. Pinafore

As you have watched the red disk of the sun sink below the western horizon, has anyone ever given you this "Believe It or Not" statement? That actually the sun has already set and you are looking at it when it is below the horizon? If at the moment this statement is made, the sun is apparently touching the horizon, then you should believe the saying, because it is true.

This apparent mystery is caused by refraction. Refraction is the bending of a ray of light as it passes from one medium to another of different density. In this case, the rays come through virtually empty space into the earth's air. Refraction increases the altitude of all heavenly bodies except those which are in the

Figure 1. *A common instance
of refraction.*

zenith, which is the highest possible altitude. The closer the body
is to the horizon, the greater the effect. Right at the horizon,
refraction causes a maximum upward shift of one-half a degree.
This angle happens to be the angular diameter of the sun and
also of the moon. Therefore after each of these has actually set, it
is raised above the horizon so that it is still visible. Conversely,
when we watch them rise on the eastern horizon they are really
still below it.

In point of fact this refraction is a great boon to us because it
increases the actual hours of daylight as much as six or more
minutes a day over the world. The amount depends on the
latitude and the season. If you work this out in terms of the extra
energy we receive, it is an enormous amount. And we owe it all to
the distorting atmosphere that surrounds the earth.

That layer of air, the atmosphere, is always between us and the
stars and planets, obstructing and distorting our view of them.
With its mixed eddies and currents, and patches of different
densities and composition, the atmosphere is rather like a veil
hiding the stars from us.

Before we look at the stars themselves, we need to know about
this veil through which we shall be looking. We shall gain an
understanding of the astronomer's trials and achievements by
considering the way we all have to peer out through an irregular,
wavering blanket, which is only semitransparent. To study the
stars or any celestial objects from the surface of the earth, we
must look through this blanket of air molecules that are jumping
around in all directions — in layers or patches of hot and cold,
sometimes murky, sometimes opaque with clouds. The higher
your elevation, the better the chance of seeing through the
blanket. And from space vehicles and the surface of the moon,
man can be free of the distortions caused by our atmosphere and
can see the heavenly bodies as they really are.

The Blue Sky

But our atmosphere has many advantages. To begin with, we are indebted to it for the beautiful blue color that we associate with a sparkling clear day. Without the atmosphere, even during the day, the "blue sky" would be permanently black except for the many pinpoints of light which are the stars, the bright planets, the enormous globe of light the sun, and its dimmer counterpart the moon. The sky always appears dark to astronauts in space. The atmosphere saves us from these black skies. For as the rays of sunlight come toward us they are affected in the atmosphere by the physical principle of "scattering." This means that the shorter wavelengths — the blue — are scattered and tossed around more than the longer — the red — and the resulting effect is of a blue sky and a yellowish sun shining in it. As the sun sinks lower toward the horizon, its rays have to penetrate greater depths in the atmosphere, which means that the longer wavelengths, orange and red, progressively begin to dominate the sky.

The sky is not the same degree of blueness all over. The sky is darkest and bluest 90° from the sun. The brightness increases very close to the sun and the color becomes whiter. The color also changes to white from the 90° zenith toward the horizon, and the light intensity increases in that manner also.

However, the outstanding Dutch scientist, M. Minnaert, pointed out that the theory that the blue sky is caused by scattered light is not completely satisfactory. It is hard to explain when, once in a great while, on an exceptional day, the sky may be seen as blue clear down to the horizon.

Twilight

Another effect of our atmosphere, and one that most people take for granted, is that of twilight. By twilight we mean both the gradual fading of daylight after sunset and the gradual brightening of the sky at the end of the night before sunrise. Twilight is the reflection of the sun's light from the atmosphere above our heads when the sun is actually below the observer's horizon. Astronomical twilight is the interval w en the sun is below the horizon but within 18° of it. Civil twiligat has a similar definition, but its limit is usually regulated to 6° for the sun below the horizon.

Every traveler soon becomes very conscious of the fact that the amount of twilight varies both with the latitude and with the season. People journeying to the Scandinavian countries, or even the northern British Isles in summer, frequently complain that they find it hard to sleep because it never gets really dark all night long. For example, at latitude 54° N. (Edmonton, Alberta) astronomical all-night twilight lasts from May 20 through July 29. From 60° N., as in Oslo and Leningrad, it lasts from April 30 through August 18. Of course from 66½° N. we are in the Arctic Circle, where on some days the sun neither rises nor sets.

In contrast to the long twilight of the regions nearer the poles, twilight in the tropics is minimal, because there the sun's path always makes a large angle with the horizon. Consequently when the sun sets, or when it rises, it covers those 18° in pretty nearly record time. There are numerous references in literature to the swift onrush of day and night in the tropics. One of the best known is by Rudyard Kipling in "The Road to Mandalay". "And the dawn comes up like thunder out of China cross the Bay."

In intermediate latitudes, as you would expect, we have no such wide variation in twilight times. In the latitude of New York, the maximum length of astronomical twilight is two hours in June, and the minimum length an hour and a half near the equinoxes.

Though everyone is aware of twilight, few persons seem to notice the beautiful phenomenon which accompanies it, the twilight bow and rise of the earth's shadow. They are best seen in the east about 15 or 20 minutes after sunset, or in the west the same amount of time before sunrise. About a quarter of an hour after the sun has set below the horizon, turn your back on it and look at the sky in the east. There you will see a dark midnight blue arch rising — the shadow of the earth. The height to which you can follow it depends on the clarity of the atmosphere, but 6°

is pretty well the maximum. At first the shadow rises at about the rate the sun is dropping below the horizon, but it gradually speeds up to double that rate.

An observer quickly learns to estimate degrees in the sky. The angular diameter of either the sun or the moon is about half a degree. At arm's length the width of the fingernail of the index finger averages 1° and the span of the open hand about 18°. Above the shadow is a beautiful arch of ethereal rainbow colors, with crimson at the lowest layer and ranging to blue in the uppermost. This is the twilight bow. It is formed by the rays of the sun, now set below the horizon, illuminating the air molecules above our heads, where only the long red rays have managed to penetrate. Higher up, the blue rays are still reaching the very rarefied air. About half an hour after sunset the twilight bow fades and the earth's shadow begins to extend upward like a curtain, the curtain of night, closing over the land.

The rising and setting of the sun or moon offer other interesting sights. One is a queer distortion in the shape. You have probably seen the setting sun look like an inverted flattened hemispherical ceiling lampshade. The distortion is caused because the refraction is much stronger right on the horizon than even half a degree above it. Refraction lifts the lower part of the sun proportionately higher than it does the upper part. Different conditions in the air layers you are looking through will provide different shapes. Fascinating series of drawings and photographs have been made by observers interested in these distorted shapes.

The Green Flash

When you are watching the sun rise or set, under rare circumstances you may see an unusual sight: a green flash. This is an intense spark of colored light, visible to the unaided eye for a few seconds just as the sun disappears below the western horizon, or before it rises in the east. The green flash is sometimes referred to as the green ray, as in this old weather proverb.

Glimpse you e'er the green ray,
Count the morrow a fine day.

Although early peoples noticed it, curiously enough it was not until Jules Verne had published his *Le Rayon Vert* in 1882 that any great interest was taken in it. In this popular book a band of intrepid tourists travels to the western bounds of darkest Scotland in search of a glimpse of the green flash, which they find novel and fascinating. Yet apparently the green flash was familiar to Egyptians early in their history, for their writings contain reference to the green color of the rising sun. Allusions to it are also found in Celtic folklore. But the earliest officially recorded observation appears to be one in 1865. There were a few more early observations, but after the publication of Jules Verne's work, observations suddenly seemed to spring up from all sides.

What is the green flash? It is a green ray from the sun left in the sky when the rays of other colors have been disposed of elsewhere by the atmosphere. At sunset, if the upper rim of the sun is nearly touching the horizon, the red light has already set below the horizon and the orange and yellow light are absorbed by the atmosphere. The blue and violet light are scattered around the sky. This leaves a narrow rim of green. Normally, the width of the green rim is too small for the unaided eye to perceive well. However, under certain conditions the width may be several times as great. Other factors increase the normal intensity too. And sometimes a weird effect results; the green flash, detached from the sun's disk, floats above the horizon after the sun has disappeared. Perhaps because Jules Verne's name was involved, for years it was suspected that the green flash was purely imaginary. There is no longer any doubt, though, that it is a real phenomenon. It has been well photographed, and observed with opera glasses and telescopes.

The longest recorded view of the green flash was that obtained in 1929 by W. C. Haines of the Byrd Antarctic Expedition at Little America, at latitude 78° S. As the sun grazed an irregular horizon of barrier ice, the green flash was seen off and on for 35 minutes. The flash was intensified by a strong temperature inversion near the ice at the time. Though Haines looked for it repeatedly under what seemed to be similar conditions, he did not get it again.

Twinkling of Stars

After the sun sets and night comes on, myriads of stars spring into view. And the phenomenon of refraction is still in force, for the stars begin to twinkle. A popular delusion is that on nights when the stars are twinkling most vividly, astronomers are working with the greatest zest. Nothing could be further from the truth. The greater the twinkling, the more seriously handicapped are observers trying to study the heavens.

The twinkling is caused by variable refraction in the atmosphere. Even though stars have diameters of from millions to hundreds of millions of miles, their enormous distance from earth means that they are seen as tiny point sources of light, even in the largest telescope. As the light from the point source passes through various layers and pockets in the earth's atmosphere, it is flung every which way. Isaac Newton noted that the twinkling of the stars was an indication that "the Air through which we look upon the Stars, is in a perpetual Tremor." This can be a very frustrating experience for an astronomer. Unless his telescope is located at a very high altitude, there are nights when the tremor is so pronounced that it may be useless to continue an observational program, even though the sky itself may be perfectly clear.

You have doubtless observed this same sort of tremor with objects on earth when you look out a window over a hot radiator and see the heat waves distort the objects or landscape you are looking at. Or you may have noticed it over a highway on a hot sunny day. This effect can be even more pronounced and lead to a mirage. Curvature of the rays in layers of hot and cold air causes this distortion. On a much larger scale, this same curvature causes the position of the star to change ever so slightly, while the rays of light that have been displaced come together in uneven amounts, sometimes making the star bright and sometimes appearing to make it fade out completely. You will notice that twinkling is more pronounced toward the horizon, where the atmosphere you look through is thicker.

A useful well-known adage is that you can distinguish a planet from a star because a planet does not twinkle. Like most adages, this is not strictly true, but it is true frequently enough to help you identify the bright planets in the sky. The planets are so close to us, relatively, that they do show a perceptible disk even though in actual diameter they are very tiny compared with stars. Therefore when a tiny ray of light from one part of this disk is deviated from the direction of our line of sight by the irregular intervening air layers, a ray from an adjoining area may take its place. The net result is a steady light. Have you ever seen the moon twinkle?

Moisture in the Atmosphere

Many beautiful effects are the results of droplets of water or crystals of ice in the atmosphere. The best known of these is the rainbow, formed by the passage of sunlight through raindrops. It frequently appears as a storm dissipates, and is taken as a promise of better things to come. The actual water droplets in which the rainbow forms are usually relatively near the watcher — a mile or two away. But this varies, and there is an actual case on record where a rainbow was seen in front of woods 10 feet away! Even so, the illusory pot of gold has never been found.

The rainbow is actually an arc that is part of a whole circle, which can be seen from a plane. The order of the colors is the one we were taught to memorize in childhood: red, orange, yellow, green, blue and violet. The relative widths and brightnesses of the colors can vary. The spherical raindrops produce a number of concentric circular arcs of different colors. The centers of the arcs are as far below the observer as the sun is above him, in angular units. The primary bow has a red outer border, while a secondary has the colors reversed. All the phenomena observed can be expressed mathematically and illustrated with elaborate diagrams.

Another frequently noticed effect of the atmosphere is a ring around the sun and, less commonly, around the moon. A ring is often taken as a warning of an approaching storm. This holds true for some, but not all, rings.

Broadly speaking, there are two different types of rings: haloes and coronas. These differ in size and coloration, and are formed in different ways. Haloes have radii of many degrees, about 22° or more, with red in the inner edge, gradually flowing into yellow, green, white and blue. The halo is caused by the refraction of sunlight or moonlight in a cloud of ice crystals, which have the shape of an hexagonal prism. The halo can indeed be a forerunner of a storm because it is often a result of cirro-stratus clouds, which are frequently storm breeders. Haloes are rather common. According to Minnaert, in some places they occur 200 days a year. Variations of the conditions that form haloes, such as the alignment of the ice crystals, also produce a host of related, less frequent apparitions such as sun dogs, light pillars and arcs.

Coronas are much smaller than haloes, usually with a radius of 1 to 5° to the brownish edge. The colors are reversed, with blue on the inside merging into yellow-white and then into the brownish outer edge. They are formed by diffraction of light in water droplets. When light passes a sharp edge it is bent slightly, a process known as diffraction. The size of the corona depends inversely on the size of the drops. Smaller drops give a larger corona.

Earthly Debris in the Atmosphere

Water droplets and ice crystals are normal constituents of the air. But sometimes large quantities of finely divided particles are hurled from the earth up into the atmosphere. This was true even before civilized man began to pollute it. For instance, the gigantic volcanic explosion of Krakatoa affected atmospheric conditions the whole world over. Krakatoa was an island between Java and Sumatra which had been built up in an old volcanic cone. The eruption occurred from August 26 to 28, 1883. A large part of the volcanic island was blown sky-high. Where once the island rose 1,400 feet above sea level, a gaping hole was left in the sea floor, 1,000 feet deep in spots. The island, previously 18 square miles in area, was reduced to seven. The actual sounds of the volcanic explosions were heard as far away as 3,000 miles.

The eruption stirred up tidal waves that reached a maximum of 50 feet in height and caused the death of more than 36,000 human beings.

According to Harry Wexler, chief of the U.S. Weather Bureau, the heavier stones hurled into the air came down on the surrounding islands to a depth sufficient to bury the forests. But the smaller ash particles were carried into the upper levels of the atmosphere and took several years to settle. They were responsible for most extraordinary atmospheric conditions the world over. At first, in the early fall of 1883, observers did not link the unusual sky conditions to the Krakatoa explosion, but as 1884 wore on, scientists began to realize the connection. And it was recalled that, just a century earlier, Benjamin Franklin had pointed out that the severe winter of 1783-84 might have been caused by the vast quantities of smoke hurled into the air by the volcanic action at Hecla in Iceland.

The Krakatoa explosion had two principal kinds of effects on the atmosphere. It produced extremely brilliant colored sunrises and sunsets. The brilliance of the red coloration was so intense that many persons mistook it for a fire. In at least one city in the eastern United States the fire department was actually called out to fight the supposed conflagration.

These colored sunsets lasted for many months, gradually diminishing in brightness. We experienced similar ones more recently, but on a lesser scale, in 1963, from the explosion of Mount Agung on the island of Bali on March 17 that year. Even in North America we were treated to many months of colored sunset phenomena with an unusual red brilliance.

A second major effect of Krakatoa was the appearance of unusually colored suns and moons which startled people all around the world. The atmospheric debris traveled around the earth from east to west in 13 days at a velocity of more than 70 miles an hour. It made two complete circuits of the earth before it dissipated. As the vast dust cloud moved, the strange colors of the sun and moon moved with it. Though it was usually described as green or blue in the tropical regions, metallic descriptions were also applied, such as coppery, silvery or leaden. The blue sun was chiefly seen at great distances from Java. The green sun was visible at first only in the Indian Ocean, but then was seen more generally than the other colors. The silvery sun was confined to a narrow zone near the equator, especially south of it. At any one place the color was not always constant during the day, but varied with altitude.

Blue moons were also seen around the world, gladdening the hearts of joke-tellers everywhere. A typical description from Centre Norfolk, England, published in January 22, 1884, states, "The moon here in November was of the intensest sapphire blue, the perfectly clear sky looking rather slaty." There seems little doubt that the well-known saying "once in a blue moon" is derived from some circumstances such as this. From a study I made of the literature of the past several centuries, it appears that volcanic eruptions and other effects will render a blue moon visible probably once in seven or eight decades. So I conclude that "once in a blue moon" is roughly equivalent in mathematical terms to "once in a lifetime."

It seems curious that extra material in the atmosphere would turn the sun or moon blue instead of making it appear redder, because the longer red rays penetrate dense air better. The explanation of the blue sun is complicated in the realm of atmospheric physics, and depends upon the sizes of the dust particles.

Hardly any people living today can remember the colored suns following Krakatoa, but many on this continent and in western Europe remember the spectacular phenomena which came as a result of smoke from forest fires in western Canada in late September, 1950. On September 24, in Richmond Hill, Ontario, I myself saw a mauve sun in the sky, shining so feebly that unless you looked for it you would not notice it. The impression of weird unreality still lingers with me.

The sun exhibited a wide range in color; it was variously described as dull red, copper, orange, greenish yellow, pink, silver, purple, blue, violet and lavender, with the last four the most popular.

The smoke came from 30 or more fires in a muskeg region of northern Alberta. After smoldering for a long time, they were fanned out of control. The damp muskeg then gave off a great volume of smoke. The smoke track was 250 to 300 miles wide, while from planes of Trans-Canada Airlines (as Air Canada was then called) the base of the smoke layer was located at 12,000 feet and the top at 20,000 to 25,000 feet. One plane that landed at Toronto airport after a flight over the Great Lakes was covered with a thin film of a brownish oily substance, and had a strong smell of wood smoke within the cabin.

The affected area extended over most of the eastern part of this continent and as far south as Florida, according to Harry Wexler. When the smoke reached the Atlantic coast it changed direction and moved northeast. It was reported over Newfoundland on September 25, and over the British Isles and Denmark on September 26. And on September 26 and 27 reports of a bluish sun and moon came also from France, Portugal, Denmark and Switzerland.

In certain places along the trajectory of the smoke, the effect on the amount of sunlight (insolation) received during the day was drastic. At Washington, D.C., more than half the expected solar radiation failed to reach the ground. At Buffalo, from 4 to 6 pm., the illumination was comparable to predawn or twilight. Sault Ste. Marie had the most spectacular drop in radiation; the amount of insolation received was less than 1 percent of normal clear-weather value! At Toronto the sudden dusk added 180,000 kilowatt hours to the demand for power. Afternoon sports events in Cleveland and Detroit required floodlights.

Over certain sections of the continent the smoke pall of 1950 provided the phenomenon known as "dark days." It was by no means the first dark day that has been recorded on this continent. Including now that of 1950, 18 dark days are documented beginning with May 12, 1706, in New England.

Dark days have been known for centuries and present an eerie and frightening phenomenon. Instead of normal daylight, there is a gradually increasing gloom until finally artificial light is needed. The unnatural darkness may last for hours or even days. These dark days are not to be confused with those in which darkness lasts a matter of minutes during the total phase of an eclipse of the sun.

The true dark days are caused by dense smoke in the

atmosphere or by volcanic dust. On this continent, forest and prairie fires have been the causes of the dark days. The outstanding dark day was Black Friday, May 19, 1780. It was preceded by gradually increasing yellowness and the typical smoky odor. Ashes of burned leaves, soot and cinders fell in some sections. A dramatic session in the Connecticut legislature when darkness became oppressive has been recorded for history by John Greenleaf Whittier in his poem *"Abraham Davenport."*

Dark days are relatives of the less spectacular phenomena of dry fogs, Indian summer and colored rains. According to F. G. Plummer, geographer, U.S. Department of Agriculture, Indian Summer is the name applied to periods during the months of October and November when the weather is unseasonably warm and there is a dry fog. A dry fog occurs when the air is saturated with impurities so fine that they are not precipitated unless a rain occurs.

Colored rains have been observed for centuries. Forest fires in some cases have doubtless been a contributing cause, as on October 16, 1785, when black rain fell in Canada or on November 6, 1819, when black rain and snow fell in Canada and the northern United States. Red rain was observed off Newfoundland in February, 1890. Rain is sometimes colored red by forest-fire smoke.

Mirages

The air can also play tricks with the way we see objects on the surface of the earth. For refraction in the earth's atmosphere causes mirages. Commonly one associates mirages with deserts, but they occur in many other locations as well, and especially near water.

In a superior mirage one or more images are seen directly above the real object. This phenomenon is common with ships at sea. The image nearest the real object is always inverted as though it were reflected from a plane mirror overhead. Unusual conditions can produce both an inverted and an upright image. In fact, there was one notable case years ago where the people of a coastal town gathered to watch a fleet of ships come in — and the ships sailed in on the sky above their heads!

The inferior mirage is more frequently seen. In this case you see images below distant objects, apparently mirrored. Most people driving on highways have seen it, for it is common on hot surfaced roads when an apparent reflection of the sky looks like water on the road. Mirages can have serious effects; for example, a mirage can mislead a weary traveler in the desert into thinking that water may be near. Even a battle between the Turks and the English in Mesopotamia on April 11, 1917, was affected by a mirage. The British commander, General Maude, reported, "The fighting had to be temporarily suspended owing to a mirage."

We have considered some of the things airborne particles do. They scatter colors around to form the blue sky. They reflect colors to form the twilight. They refract colors to make rings around the sun and moon, and mirages. Refraction in the air can make things which are on the ground appear up in the sky, and things which are in the sky appear on the ground. Is it any wonder that people report flying saucers?

Flying Saucers

Flying saucers may properly be considered in any one of several sections of this book, for there is definitely a wide variety of objects that have been classed as flying saucers. Some arise from the kind of atmospheric effects with which this chapter deals. Some of the bright objects existing outside our atmosphere, the bright planets, especially Venus, have been mistaken for flying saucers — even by airline pilots. And some of the events in our atmosphere discussed in the next chapter have been considered as flying saucers.

Into the turbulent rotating air masses on which forces of heat and cold and gravitation are already operating on a grand scale, man injects his own materials. Smoke and noxious gases, balloons, flying machines and secret experimental craft with their exhaust gases, rockets, even electrical currents leaking from powerful transmission lines — all these and more help to complicate our air blanket.

In addition to this turmoil in our air there is an important psychological factor for flying saucers. From time immemorial men have seen things in the heavens that were not really there. Delusions or hallucinations or whatever you call them, no one can deny they exist. Some of the early descriptions of comets, quoted in the chapter on them, show the great gap between reality and the recorded description.

Then add to this mixture the fact that since flying saucers became one of the most popular topics of the great North American continent, they have become an invitation to ingenious do-it-yourselfers to construct some of their own to delude and entertain the populace.

With a mixture of probabilities and possibilities like this, it is no wonder that there are reports of flying saucers. The wonder is that there are not more!

Official committees have investigated reports of the flying saucer phenomena, and members of these committees have not always areed with the verdict. Several conclusions seem to have emerged from the studies, however. First, the great majority of the sightings can be explained by natural causes, some of which we have mentioned. Secondly, there is a residue of sightings by a number of creditable witnesses for which the explanation is not yet at hand. However, this flying saucer residue is but one of many natural activities still unexplained, as discussed by Dr. Peter M. Millman, with examples, in *The Journal of The Royal Astronomical Society of Canada,* August 1975. Our knowledge continues to grow and while now we see through a glass darkly, in the future we will see face to face. Consider the enormous progress man has made in the past hundred years in understanding the physical laws of the earth and of the whole universe.

There is still no physical evidence that can be produced and held in the hand that these flying saucers have come from beyond the earth. Contrast this fact with meteorites, which without a doubt do come from beyond the earth. Scattered in museums and private collections around the world are thousands of meteoritic specimens. These can be proved by chemical analysis to resemble no rocks on earth, and to have been formed under conditions that our earth could never produce. Furthermore, every year dozens of bright fireballs of the type that drop stones are seen over an area of thousands of square miles and viewed by thousands of people. Contrast this with the flying saucer sightings, which are seen over a very limited area, sometimes only fractions of a mile, and usually by relatively few people.

Nor have flying saucers been tracked over really large distances by telescopes or radar. With the satellite-tracking systems now encircling the earth, with all the radar units operating on the surface of the earth, it is almost impossible to believe that a space ship could arrive from a distant planet without being detected somewhere along the way by many instruments in this fast tracking network. If our technology is sophisticated enough to follow space probes tens of millions of miles from earth, surely it could pick up an unknown, sizable object that makes a trip in and out of the earth's atmosphere.

However, we wish to make it clear that just because some of us scientists believe, as of this writing, that there is no scientific evidence for the arrival of a flying saucer from beyond the earth, we certainly would not suggest that it is impossible that someday one will arrive. It could happen any time. What a wonderfully exciting day for the human race!

There are two subjects that often lead to heated debate if an astronomer tries to discuss them with laymen. One subject is flying saucers; the other, astrology, the idea that people's lives are affected by the zodiacal signs in the heavens. Astronomers do not support this theory. Usually neither side is receptive to arguments advanced by the other, and the discussion ends in a deadlock. So having introduced these two subjects, we leave it to the reader to pursue them further in other books if he wishes.

★

TWO

Natural Events in the Earth's Atmosphere

There are more things in heaven and earth, Horatio
Than are dreamt of in your philosophy.

Hamlet I, v.

While our atmosphere is affecting the appearance of objects we see through it, as discussed in the last chapter, events are happening within it which we can observe from the surface of the earth. These events are caused by material that enters our atmosphere from above, from sources beyond the earth.

Aurorae

Aurorae form one category of such happenings. The aurora

borealis, or Northern Lights, has a counterpart in the aurora australis, the Southern Lights, in the southern hemisphere. Together, they are called the Polar Lights.

The flickering colored draperies or arcs of the aurora always cause wonderment in the beholder. Auroral expert Sydney Chapman said that many superlatives can be justly applied to the aurora: "grand, majestic, marvellous, mysterious, infinitely various and changing".

Unfortunately, only a small percentage of mankind ever sees the aurora. For millions living in the tropics it is virtually unknown. The northern part of the United States and Canada are in a favored position for the viewing of the Northern Lights. Elias Loomis of Yale apparently was the first to find that the aurora became stronger and more frequent as one went northward but that this did not continue right up to the north pole. In 1860 he provided a map of the auroral zone, which is centered, not on the geographical pole, but rather 10° away from it, the north geomagnetic pole. A map of the lines of equal frequency of auroral visibility over North America shows that across the northern United States on the average a couple of dozen aurorae may be seen each year. This number jumps to 100 at Nome, Alaska, and the tip of James Bay in Canada. It falls down to one a year on a curve connecting New Orleans and Los Angeles. Only one a decade may be seen at the latitude of Mexico City. Records show that in two centuries, 1730 to 1921, 12 very strong displays permeated to very low latitudes.

Aurorae are much farther up in our atmosphere than the sources of the apparitions discussed in the first chapter. They are in the ionosphere above the weather region, with 99 percent of the atmosphere below them. Aurorae that occur in the shadow of the earth lie from 55 to 150 miles above it. Aurorae that shine in the sunlit air to the west after sunset or to the east before sunrise may extend to very great heights indeed, up to 500 miles or so.

Because an aurora is such a high-level phenomenon, the same one can be observed over a large area, hundreds or even thousands of square miles, depending upon its height and strength. An aurora has a thin ribbonlike structure. While in height it is the order of 100 or more miles and may extend literally thousands of miles, its thickness is only two or three miles. And certain types of aurorae — the rayed, for example — may be only a few hundred yards thick.

Although the aurora has been known and mentioned for millennia, only in the past 100 years or so have regular scientific studies been made of it. Even now, in the sections where it can be seen, the understanding and explanation of the aurora are by no means as widespread as its apparition. A common myth that keeps cropping up is that the aurora is caused by sunlight reflected from icebergs.

Actually the sun *is* responsible, but not in this way. The aurora is triggered by particles from the solar wind. "Solar wind" is a new term in astronomy the last couple of decades. It applies to the stream of high-energy particles pouring out from the sun into the regions of the planets, especially after a solar disturbance such as a flare has unleashed it. The cloud of solar gas takes about a day to travel from the sun to earth, with a mean speed of 1,000 miles a second.

These high-energy particles become entrapped in the earth's magnetic field, following certain prescribed paths. They end their journey by activating a display of aurorae something akin to turning on the switch for a neon light.

One of the amazing facts about the aurora is its complete disappearance from England during the 17th century, although several displays were seen on the continent of Europe during this interval. On November 14 and 15, 1574, Stowe recounts that "the heavens from all parts did seem to burn marvellous ragingly." The next occurrence in England was not until March 17, 1716. In all this interval of 142 years the heavens were keenly watched by the Baconian philosophers of the Royal Society.

On the continent the aurora was seen as late in the century as 1629. In Chile the Southern Lights were very strong in 1640. But from then on the Polar Lights died out everywhere, though they were seen again in northern Scotland in 1691. Even in the high latitudes of Iceland and Norway they became so rare as to be considered omens.

The return of the aurora caused a great stir in one place after another. In Copenhagen in 1709 the guard turned out with arms to prepare for some unknown but imminent catastrophe. In England, Halley compared its appearance to the concave of the great cupola of St. Paul's church. The almanacs referred to it as this "Great Amazing Light in the North." In the lowlands of Scotland for a long while the aurora was known as "Lord Derwentwater's Lights" because it occurred on the eve of the execution of the rebel lord on February 23, 1716. At Bologna the lights were supposed to have been unknown until 1723. Appearances in China in 1718, 1719 and 1722 created such a stir that engravings were made by thousands. These had to be distributed secretly since portents were contraband inside the Great Wall.

Now we know that auroral and magnetic activity on earth are correlated with the sunspot cycle, discussed in Chapter 4. The peaks in this activity follow a year or two after the 11-year peak in sunspot numbers. In high latitudes this correlation shows up more in the brilliance and intensity of aurorae, but in lower latitudes it is evidenced by the frequency and number of their appearance.

A controversial aspect of the aurora is its audibility. Many people have from time to time insisted that they heard the swishing of its draperies. And yet the aurora is 50 miles or more from the observer. At the speed of sound — five seconds to go one mile — the noise would be expected to lag noticeably after the visible flashes. Eyewitness accounts from those who have heard the aurora are nevertheless very convincing. Wordsworth refers to it in his poem *"The Complaint of a Forsaken Indian Woman"*: "In sleep I heard the northern gleams. . ."

A very strong aurora and exceptional sensitivity in the beholder seem necessary ingredients for reports of audibility. The phenomenon is mainly confined to very high latitudes. Northern Canada is a favored location. In fact, in the very first issue of the *Journal* of the Royal Astronomical Society of Canada in 1907 the problem was raised, and thereafter many "ear witness" reports were published. William Ogilvie, a surveyor-explorer of the Yukon Territory, never heard the aurora, but described one member of his party who consistently heard slight rustling sounds. During a bright aurora in 1882 the man was blindfolded. At almost every brilliant rush of auroral light he exclaimed, "Don't you hear it?"

In 1911 Mrs. George Craig, amateur astronomer and wife of a Department of Justice official in Dawson, Yukon Territory, heard the aurora for the first time. At 1:30 A.M. on January 26 she and her husband, returning from an evening out, were crossing the park in the still of the night. As Mrs. Craig remarks, "There are no horses and rigs moving about at fifty below zero at midnight in the Yukon. Nothing noisier than the soft-footed Malamuit *[sic]* dog."

"We were arrested by strange sounds, like the swishing and brushing together of particles of finely-broken glass. The sound came in great waves, passing slowly backwards and forwards over the auroral arc. Sometimes the wave, with its musical tinkling, would almost seem to surround us; then it would recede so far as to be almost inaudible."

It would appear incontrovertible that on rare occasions, and to certain ears, the aurora does make audible sounds. I once asked an Inuit, George Koneak, from Fort Chimo, Ungava, if he had ever heard the Northern Lights. His detailed reply firmly convinced me that on occasion they are audible. "Oh yes," said

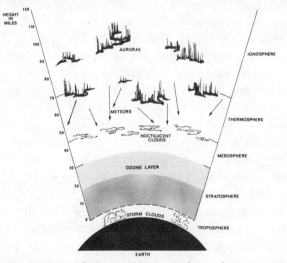

Figure 2. *Cross-section of the earth's atmosphere, showing regions where meteors and aurorae commonly appear.*

George. "One day I was trying very hard to get home with my dog team, and I figured if I traveled all day and into the night I could get home by midnight. Along about nine o'clock at night — the sky was very clear and the stars, my, how they were shining — and the Northern Lights were very brilliant and then I began to hear them. They went whoo-o-o-sh, whish- whoo-o-o-sh, like that. It was not wind. The night was very calm. All was still. But the Northern Lights they went whooo-o-sh, and the dogs were scared. They scattered in all directions they were so scared, and my friend and I, we had hard work getting the team together again."

As scientists learn more about the aurora, the reasons for its rare audibility will doubtless become clearer.

Shooting Stars

Anyone who has spent time out of doors after dark, away from the heart of a big city, has probably seen a shooting star. Shooting stars, or falling stars, are technically known as meteors. They are one of the most commonly observed of all astronomical phenomena.

Meteors first began to be understood in 1798, when their heights were measured by Brandes and Benzenberg, two young students at Göttingen who observed meteors simultaneously. To measure heights you need at least two observers separated by a known distance, say 20 miles. The two can see the same meteor simultaneously projected against a different background of stars and constellations. Since the meteor is vastly nearer us than are the stars, the observers will see it appear to shift with respect to them. The average meteor becomes visible about 60 miles up and fades out around 40. The larger the meteor, the greater the height at which it starts to glow.

The luminosity of a shooting star comes from the destruction of its mass as it collides with the earth's atmosphere. The death of a shooting star is a brilliant phenomenon. As the mass is consumed, the energy that the little meteor carries is transformed into light. And the light produced depends on the *square* of the speed with which it is traveling. For shooting stars this speed can be very great — up to about 48 miles a second — far exceeding even the speeds produced nowadays by man in satellite or space-probe launchings. A meteor traveling at 20 miles a second can, under the same conditions, produce four times as much luminous energy as one moving at 10 miles per second.

Consequently, the bright streak you see going across the sky may actually be caused by something not much larger than a dust speck. Its fleeting brilliance is so great that you may not even notice the real stars alongside — pinpoints of light, but each one actually a huge body composed of as much material as maybe half a million earths! Shooting stars serve to bring into sharp focus the part "proportion" plays in our understanding of the heavens.

A question frequently asked is, "What would happen if a real star fell?" And the answer is simple. The real stars *are* falling all the time, in the sense that they are all moving with sizable velocities. These are measured from a few miles a second up to more than 100 miles a second for a few high-speed or runaway stars. But so great are the distances separating our solar system from even the nearest stars that, despite these great velocities, there is no star in the sky whose motion you will be able to detect with the unaided eye in a human lifetime. But if a person who knows the constellations well now could come back to earth in 1,000 years, he would be able to see that several of the bright stars have shifted their positions slightly, by amounts comparable with the apparent diameter of the moon.

It is fun to stand out for an hour or two on a clear night and watch meteors dart across the sky. On an average you should be able to see one meteor every quarter of an hour in the early evening in the spring. However, if you are up before dawn in the autumn this will increase to one every four minutes because you are then riding on the forward side of the earth as it rotates and revolves around the sun. So you get what can be described literally as a bumper crop.

Estimates of the number of shooting stars hitting the earth's atmosphere every day run into astronomical figures. The number that could be seen is probably 100 million a day. Undoubtedly a large proportion of this material originally was in comets, and we are being treated to the disintegrated fragments of their tails.

The Leonid Showers

For some decades after the young Göttingen students first put the study of shooting stars on a scientific basis, the interest in these objects was rather desultory. Then an event occurred in the middle of the night that quite literally forced the world to wake up and take notice. This was the great shower of Leonid meteors on the night of November 12, 1833, which reached its greatest brilliance in North America. It caused a stir that can never be forgotten, and may have triggered some of the religious sects that sprang up soon after, because of the fear generated that night that the world was coming to an end. It forced the scientific men of this continent to begin the study of meteors in earnest.

When you watch meteors, mostly their trails will seem quite haphazard to you, crisscrossing the heavens in any direction and from any direction. On some occasions, however, the trails make a definite pattern. Each trail projected backward will seem to come from the same point among the stars as those that flew by earlier. (There will always be a few trails of sporadic meteors.) The tiny region of the sky from which the meteor trails seem to emanate is called the radiant. The meteors that form this pattern are named after the constellation in which the radiant lies, as the Leonids from Leo and the Perseids from Perseus. These meteors are going around the sun in a swarm, and when the earth intersects them in their orbit we have a shower of shooting stars.

The Leonids are one of the greatest showers of meteors. Actually the shower of 1833 was a repetition of one that had appeared in 1799 over much of North and South America. Probably because of the sparsity of North American population then, the 1799 shower did not attract as much attention as that of 1833. The 1799 shower was described by Andrew Eliot who was sailing from Philadelphia to New Orleans, and was called up at 3 A.M. on November 12 to see the "shooting of the stars." "The phenomenon was grand and awful, the whole heavens appeared as if illuminated by skyrockets, which disappeared only by the light of the sun after daybreak. The meteors, which appeared at any one instant as numerous as the stars, flew in all possible directions, except from the earth toward which they all inclined

Figure 3. *The Leonid shower of 1833 as seen over Niagara Falls. (From* Harper's Magazine, *Vol. I, 1855.)*

more or less; and some of them descended perpendicularly over the vessel we were in, so that I was in constant expectation of their falling among us."

November 13, 1833, lives in our legends as "the night the stars fell on Alabama." This shower was witnessed by thousands of persons in North America. The sky was described as like a giant umbrella with flaming spokes. A famous drawing in the first volume of *Harper's Magazine* shows the Leonids falling over Niagara Falls. A careful and colorful description, published in the *Columbian Centinel,* comes from an anonymous observer at Boston, who "arose as usual at four o'clock." At 5:40 A.M. when he opened the parlor shutters, he saw innumerable meteors. "They were moving in a direction downward, and were falling about half as fast as the flakes of snow in one of our common snow falls, with intervals of a few seconds, when there was not so many." In the 15 minutes before six o'clock he counted 650 falling stars in the portion of the sky he could see, and estimated 8,660 for the sky as a whole. This made a rate of 34,640 an hour, according to Denison Olmstead, of Yale College, who summarized all reports, with about 100,000 in the three-hour interval of the shower.

When a meteor is brighter than any star or planet, it may be called a fireball. Some fireballs accompanied the millions of shooting stars in the 1833 shower. Alexander Twining of Yale described a fiery ball of deep red falling vertically down from an altitude of 20° to the tops of hills at 4°, "as if a column of glowing melted metal had been poured down from the spot whence the meteor issued."

Many witnesses, however, were more impressed with fear than with beauty. A planter in South Carolina described the effect of the shower on his plantation workers.

"I was suddenly awakened by the most distressing cries that ever fell on my ears. Shrieks of horror and cries for mercy I could hear from most of the negroes of the 3 plantations, amounting in all to about 600 or 800. While earnestly listening for the cause I heard a faint voice near the door, calling my name. I arose, and, taking my sword, stood at the door. At this moment I heard the same voice still beseeching me to rise, and saying, 'O my God, the world is on fire!' I then opened the door, and it is difficult to say which excited me the most — the awfulness of the scene, or the distressed cries of the negroes. Upwards of 100 lay prostrate on the ground — some speechless and some with the bitterest cries, but with their hands raised, imploring God to save the world and them. The scene was truly awful; for never did rain fall much thicker than the meteors fell towards the earth; east, west, north, and south, it was the same."

In 1866 the shower was again repeated on November 13, but this time it was at its height in the British Isles, where at Greenwich eight observers counted 8,000 with 4,860 in one hour. On November 13-14 of the following years, observers at the Magnetic Observatory in Toronto recorded 2,286 meteors in 1867 and 2,486 in 1868. In 1899 only a sprinkling of Leonids appeared, a great disappointment, and again in 1933. As Henry Norris Russell said when it became obvious that the Leonid swarm had been pulled from its course by planetary perturbations, "An overpass has been substituted for a grade crossing."

In 1966, to the joy of those who saw them, the Leonids put on a superb performance again just before dawn, particularly at high altitudes in the southwestern United States. At Kitt Peak near Tucson, Arizona, the rate increased from 30 a minute for a single observer through several hundred a minute to a maximum 40 per second. Just before dawn the rate fell to 30 per minute. Observers commented that looking directly at the radiant gave them an impression of the earth moving through space toward Leo, so steadily were the meteors coming from that direction.

Meteor showers are directly connected with comets. They are composed of the material which the comet has shed in its journey round the sun. The Leonid orbit is linked with that of Temple's Comet (1866I), which has a period of 33 years. The Leonids form a swarm of material in their orbit — and every 33 years the earth crosses near the swarm. Perhaps we will have another spectacular Leonid shower toward the end of the century.

The Draconids or Giacobinids, associated with Comet Giacobini, gave splendid showers in 1933 and 1946 and may reappear again.

Annual Meteor Showers

Meanwhile there are a number of less spectacular but still interesting and dependable annual showers. When the material in a meteor swarm is spread out along its orbit, and not bunched together, then we may have an annual shower. There are ten such showers that meteor observers watch for each year. *The Observer's Handbook* gives the dates of these showers.

The Perseids are the best known and the best observed of these. They visit us in August, when warm summer nights and holidays beckon people out of doors. The annual number of Perseids has been relatively constant for the past hundred years. At the peak of the shower (around August 12) a single observer may see 50 meteors an hour. Because the Perseids have a high velocity of 37 miles per second, they are brilliant meteors and about half of them leave trails. An average Perseid of 2nd magnitude has the mass of a few milligrams, a pinhead. A similar meteor traveling more slowly, such as a Draconid with a velocity of four miles a second, would be several magnitudes fainter — about the 5th magnitude. The Perseids begin to appear at a height of 70 miles. The trails occur in the region of the atmosphere above 50 miles, where the temperature reaches its minimum of −70°C.

The material forming these meteors is spaced uniformly around a highly elongated orbit at right angles to the plane in which the earth is moving, and extending out about 23 times the earth's distance from the sun. Probably these Perseids have been in the same path for thousands of years. Each Perseid meteor weighing less than .03 oz. is the largest occupant of a space of 6 million cubic miles. On the average the particles are separated by over 100 miles. Fletcher Watson of Harvard estimates the total mass of Perseids to be 500 million tons, which would make a layer nearly one inch thick over the state of Connecticut. The orbit of the Perseids has been found to be the same as that of Comet 1862 III, Tuttle's. This comet has a period over 100 years, and is expected to return toward the end of the century.

Meteoritic Dust

As for the little dust specks that are falling, the earth is under constant bombardment from them. We are indeed fortunate that our atmosphere is acting as a meteor bumper and fending off most of the interplanetary particles before they hit the ground — or us. Close-up photographs of the lunar surface show what happens when there is no atmosphere to protect it. Space ships without a favorable atmosphere surrounding them will have to provide their own meteorite bumpers. There was concern when Skylab I lost its shield.

Eventually most of the meteoritic dust settles down to the earth's surface. It is so fine that it takes many weeks to float gently down inconspicuously. The total mass may amount to as much as a thousand tons a day. This settled dust is a much sought-after commodity, which scientists are busy trying to collect. They hunt in arctic and antarctic wastes, where it is relatively uncontaminated by terrestrial dust, and in sediments from the bottom of the ocean, where it was deposited over long periods of geological time.

Sometimes the dust in the air seems to precipitate cloud

formation. Correlations have been noted between rainfall and large amounts of meteoritic dust in the atmosphere. Then a rare type of cloud known as "noctilucent," very high in the atmosphere, seems to be connected with the dust. The name comes from the fact that it shines when the sun is well below the horizon. These are sometimes called mother-of-pearl clouds because of their iridescence. Of great beauty, in form they may be veils or sheets, bands, crests and whirls. Their colors range from bluish tints to gold near the horizon, with other delicate rainbow colors. Somewhat surprisingly, observers have not mentioned the greenish shades characteristic of the aurora. They are mostly formed about 50 miles above the earth's surface, where strong upward currents come to a stop. The air is about 100,000 times thinner than at the surface.

And they are strongly limited in the places and times they can be seen. They have been observed only between latitudes of 45° to 77°, usually in much narrower limits, between 50° and 65°. Also, they have been seen only when the sun is between 6° and 16° below the horizon. Reports of noctilucent clouds from the southern hemisphere are rare, but increasing in number, as efforts are being made to establish a network of observers there. Unfortunately, the antarctic bases are outside the optimum latitude range.

Noctilucent clouds were seen more commonly from 1885 to 1894, and first recognized by a German scientist, O. Jesse, as a particular variety of cloud at that time. It was not meteoritic dust, but dust from the great Krakatoa explosion, which produced them then. Effects similar to noctilucent clouds have been seen after rocket firings.

In 1962 those tiny dust particles, only 1/100,000 inch in diameter, were actually gathered by rocket-sampling experiments in Sweden. They were dust nuclei with traces of nickel and iron, showing their meteoric origin, and coated with some volatile material, probably ice. The particle density at this region of the atmosphere (the mesopause) was found to be 100 to 1,000 times greater when noctilucent clouds are seen than when they are absent.

On May 19, 1910, the day after the earth went through the tail of Halley's comet, unusual atmospheric effects were seen over North America. At the Yerkes Observatory Dr. Frederick Slocum photographed some remarkable clouds, between noon and 1 P.M., like mother-of-pearl clouds. The Observatory staff watched the display for an hour — like nothing they had ever seen before. Accounts of unusual phenomena that day were collected by Professor W.J. Humphreys. In addition to the strange clouds, they included haloes, coronas and Bishop's rings, a reddish-brown corona around the sun first noted after the Krakatoa explosion by S.E. Bishop of Honolulu. The dust from the comet's tail was doubtless responsible.

Meteoric particles still in space, which have not yet entered the earth's atmosphere to become meteors, are called meteoroids. Because of the serious hazard that meteoroids may pose to interplanetary space vehicles, artificial meteors are being created to bring us more information about the real meteors, such as their masses. Many such experiments are being carried out by the Meteor Simulation Project of Langley Research Center to determine the relation between brightness of a meteor and its mass. For example, a rocket Trailblazer II vehicle was launched from Wallops Island, Virginia, on the night of March 17, 1966. An iron pellet, only 1/5 ounce in weight, was boosted to an altitude of approximately 180 miles. The pellet re-entered the atmosphere 383.7 seconds after the launch, with a velocity of almost seven miles a second. Off the coast of North Carolina the visible streak it produced in the sky lasted for one second. For observers within 60 miles of the meteor it was nearly as bright as the brightest star in the heavens. The meteor began to glow at a height of 42 miles, and was seen down to 36 miles. When the iron pellet entered the earth's atmosphere at seven miles per second it converted 0.4 percent of its kinetic energy into light, most of which came from atomic radiation of the vaporized iron.

Thus far we have been discussing mainly meteors that resemble stars in their brightness as they streak across the sky. But some people have the good fortune to see a much larger and more brilliant object passing across the heavens — something as bright and sizable as a piece of the moon sailing by, perhaps accompanied by loud detonations. This is a fireball or bolide.

A fireball is one of the most spectacular of celestial objects. Its brilliant light may actually light up the whole sky. The Pribram, which fell in Czechoslovakia on April 7, 1959, had a magnitude of −19, that is, a brightness 300 times that of the full moon. The trail from such a flaming object coursing the skies may last for many minutes and assume curlicue shapes as the current aloft causes it to drift. Fireballs can be seen over an area of hundreds or even thousands of square miles. Frequently a fireball is accompanied by loud detonations as the shock wave from its rapid passage traverses the atmosphere to earth. A fireball of November 23, 1877, over the British Isles caused a roar that was 100 times greater than a thunderclap. Those who hear a detonating fireball are nearer to it than those who merely see it gliding through the skies.

Many fireballs drop stones to earth. This has been known from very early times, and some of the resulting meteorites have been carefully preserved, and even worshiped. In Egypt, artifacts worked in meteoritic iron have been found from the prehistoric period, before 3100 B.C. Later the term "the iron of heaven" was used. The oldest descriptions of actual meteoritic falls are from the Chinese, dating back to 644 B.C. Probably the "image that fell down from Jupiter," worshipped in the temple of Diana at Ephesus, was a meteorite. At Mecca a black stone that fell was built into a corner of the Kaaba in the 7th century A.D. Believed by the Moslems to have been given to Abraham by the angel Gabriel, it has been of tremendous influence. So sacred is the stone that it has never been submitted to actual test to prove that it is a meteorite.

During medieval times a number of historic, well-documented stones fell on Europe. About 1400 A.D., an iron meteorite weighing 235 pounds fell in Bohemia and was preserved in the Rathaus of Elbogen. The politicians interpreted it as the metamorphosed remains of a previous cruel ruler. On November 16, 1492, a 280-pound stone fell at Ensisheim in Alsace and has ever since been displayed in the church there.

In Canada a meteorite at Iron Creek, Battle River, north Alberta, was worshiped by the Cree and Blackfoot Indians. Members of these tribes made pilgrimages to the top of a hill in

Alberta where the "Manitou" stone rested. The Indians called it "Pe-wah-bisk Kah-ah-pit" (the iron where it lay). This meteorite weighs 386 pounds and is 91 percent iron, the rest mainly nickel, and is now in the Provincial Museum of Alberta in Edmonton. It is one of many found and revered by the Indians of North America. Since these Indians had no written records, almost nothing is known of the falls or the times they occurred. When archeologists uncovered a Montezuma Indian ruin in Mexico they found the Casas Grandas meteorite wrapped like a mummy.

The curious thing is that, by the 18th century, well-informed men refused to believe that stones could fall from the sky. This paradox is difficult to explain, but it seems to be an illustration of the proverb that a little learning is a dangerous thing. Once man began to have glimmers of understanding of the universe about him — and especially the vast distances separating the earth from the other bodies — then, despite the centuries of documented falls, he refused to believe that stones could fall out of the blue vault of heaven above us.

This strange prejudice reached a sort of zenith with a fall of stones in Connecticut in 1807. When U.S. President Thomas Jefferson was told of it, he replied, "I would sooner believe that a Yankee professor would lie, than that a stone would fall from heaven." However, on the other side of the ocean, an earlier fall on April 26, 1803, at L'Aigle in France, about 80 miles west of Paris, had been witnessed by many inhabitants. At one o'clock in the afternoon, between 2,000 and 3,000 stones fell to earth. Such an experience could hardly be written off as just imagination! Some years earlier, the scientist E.F.F. Chladni had detected a difference in chemical composition between terrestrial rocks and some stones said to have come from the sky. So evidence began to pile up to confirm that stones do indeed fall out of the sky. The study of meteorites was put on a scientific basis.

Now we know that all meteorites are fireballs as they pass through the atmosphere, but not all fireballs drop meteorites. The ability to survive a passage through the atmosphere will depend on the mass, composition and speed of the meteoroid. Some meteors may consist of ices with small amounts of dust, like comets, and are called cometoids. Their brilliance may be due to an intrinsic luminescence, from the heating of the substances they contain, such as nitrogen, which may account for the green light of some slow-moving fireballs. The fireballs of the Leonid showers dropped no stones, and were perhaps of this nature.

Meteorites come in the slowest range of meteor velocities. If the meteor is traveling faster than 14 miles a second when it hits the top of the atmosphere, not much of it can survive. The chances of survival can be known by determining its speed when the light goes out at an altitude of about 12 miles. If this speed is between two and three miles a second, the chances are good that one or more meteorites reach the ground. If the speed is six miles a second or greater, there is little chance that a sizable meteorite mass will survive. The last few miles of its path the meteor is going too slowly to be luminous. It is never burning when it hits the ground, and seldom hot.

Meteorites are also classified according to their chemical composition, in two broad groups of stones and irons, with many gradations within each group and between them. Commonest in space are the stony meteorites (aerolites), which comprise more than 90 percent of all observed falls. Those freshly fallen have a black crust, sometimes with a glassy veneer. When weathered, they are hard to recognize from terrestrial rocks. About 10 percent of the falls are metallic meteorites (siderites), alloys of iron, nickel and cobalt. In a chemical etching process they show a characteristic lattice known as Widmanstätten figures, one of the clearest tests proving meteoric origin. A rare group representing only 4 percent of all meteorites is the siderolites, transitional between stones and irons.

The stony meteorites are classified as chondrites and achondrites. The term "chondrite" comes from the small spherical grains that characterize them. Stony meteorites without these grains are called achondrites. The chondrites are by far the commonest type of meteorite in space.

Falls and Finds

Meteorites are classified as to being "falls" or "finds." If the meteorite is seen to come down, it is a fall. An estimated 500 meteorite falls occur each year over the whole earth, but 300 of them go into water. Finds are those which are picked up years, decades or centuries later when the stone on the ground is recognized as unusual. Most of the known meteorites are finds rather than falls. Among the finds nickel-iron meteorites predominate. Dr. Peter Millman of National Research Council of Canada has catalogued meteorites recovered from 41 different arrivals in Canada. Only about a quarter of these are falls. In recent years, though, the proportion of falls is rising as press cooperation and the public help to locate the end point of bright fireballs. Many meteorites must be lying undetected on Canadian soil. Rewards are offered by National Research Council of Canada for new meteorites. A postal slogan on some of the correspondence from their offices reads:

REPORT   SEND
BRIGHT   SUSPECTED
METEORS   METEORITES

The largest meteorite fall yet witnessed this century was the Sikhote-Alin in eastern Siberia at 10:35 A.M. local time on February 12, 1947. At longitude 134°E., latitude 46°N., this is near the Sikhote-Alin mountain range, and near Ussuri, after which the fall has also been named. The total mass of meteoritic material that reached the earth was about 100 tons, of which five tons were collected by early expeditions.

No scientist from the outside world was there at the time, and reports of the fall are somewhat conflicting. An oblong fiery body with a multicolored tail was seen by villagers over a radius of more than 60 miles. For about six seconds it passed through the sky from north to south and then disappeared. Since this was near noontime, the light was not so spectacular as for a nighttime fall. When about six miles above the earth's surface, the body was severely retarded by the braking effect of the earth's atmosphere and disintegrated into thousands of fragments. A few minutes later thuds were heard as far as 120 miles from the main place of fall.

The well-known Soviet astronomer, E. L. Krinov, who investigated the happening, described it as an entirely unique rain of witnessed fall. There was no explosion — the solid mass was not suddenly transformed into gas. Instead, the heavy shower of metallic meteorites came down on a strewn field in the shape of an ellipse. The major axis was about three miles long, and the area about four or five square miles, divided into a head and tail region. The larger masses fell in the head area where all the craters are. Near the largest crater, 95 feet in diameter, a forest of 100-year-old cedars vanished without a trace. Around small craters the individual trees which remained standing were shot through with fragments. Hundreds of specimens were collected in the tail region. Most of them were lying on top of the forest carpet, not buried in soil.

A well-observed fall of a gray chondrite was that at Bruderheim, in Alberta, Canada, on March 4, 1960. The meteorite passed over the large city of Edmonton at 1:06 A.M., just as the late TV show was ending. People closing windows and doors for the night had an unexpected treat.

Reports, gathered and analyzed, show that the fireball flared a bright blue-white at a height of about 30 miles. It held this illumination for 25 air miles until it detonated at a height of 16 to 17 miles. Bright fragments were thrown off at the detonation point. The flash was visible for 200 miles, and the detonations gave a sonic boom heard over an area of 2,000 square miles. The fireball's velocity was eight to 10 miles a second. The total light of the meteor could be estimated from an all-sky camera operating at a meteor station at Newbrook, Alberta, which gave a value of −19, halfway between the sun and moon in magnitude.

The fireball was first seen west of Edmonton at Duffield, Alberta, by Alexis Simon of the Paul's Band Indian Reserve. He saw it light up the sky as it passed from the northwest to the northeast giving off flashes of fire. He also heard a rushing sound, like a high wind, which lasted for five to six seconds after the fireball passed.

The first stone of the fall to be picked up was found by Nick Broda, a Bruderheim farmer, in his barnyard on the day of the fall. It was identified as a meteorite at the Sheritt Gordon Nickel Refinery at Fort Saskatchewan. This spurred a systematic hunt of the region by scientists associated with the University of Alberta, which acquired most of the specimens.

About 200 sizable fragments were gathered, some from the ice of the North Saskatchewan River. They fell in a well-defined ellipse, three and a half miles long and two and a quarter miles wide. The larger individuals, which carried farther in their flight, were near the southeast apex. In total weight this is Canada's largest meteorite, at several hundred pounds. The scientists recovering the meteorites were careful to make precise records for each piece, where it fell, whether it made a little crater and what it weighed. Most of the pieces weighed more than six pounds each, with a preponderance of 50 to 60 pounds. This seems to be the optimum size for this kind of gray chondritic meteorite to maintain as it penetrates the earth's atmosphere at a speed of eight miles a second. This wonderful rain of meteoritic material did no damage, but presented scientisits with many pounds of fresh extraterrestrial matter from outer space.

The meteorites known as the carbonaceous chondrites are the rarest type, and among the most intriguing. This stony type contains carbon, hydrocarbons and water, and very little metal. A spectacular fall of these stones came near the Mexican town of Pueblito de Allende in southern Chihuahua on February 8, 1969, at 1:05 A.M. (C.S.T.). This fall increased by many fold the amount of carbonaceous chondritic material available for analysis in the world. Up to two tons of material may have come down as a brilliant blue-white fireball moved across the sky and the ground began to tremble. Analysis at the Smithsonian Astrophysical Observatory showed that fragments of the Allende meteorite are perhaps the oldest rocks known. With an age of about 4.5 billion years, they are older than the oldest moon rocks examined to date, but not older than the lunar dust. Nowadays the recovery of meteorites shortly after arrival on earth, which permits immediate analysis, is leading to much greater knowledge of their history and make-up.

Three countries have established networks for meteorite recovery, Canada, Czechoslovakia and the United States. Batteries of cameras, "electronic bloodhounds," search the skies to record fireball trails precisely enough to find the endpoint if the fireball drops a stone. The recovery of meteorites has been much less than expected, but very valuable data have been acquired (as already quoted) on the disappearance of fireballs.

The program sponsored by the National Research Council of Canada is known as MORP, the Meteorite Observation and Recovery Project. A network of 12 tracking stations was established in 1970-71 on the prairies to aid in rapid recovery of meteorites by photographically recording their fall to earth from at least two stations. The tracking stations are pentagonal observatories spaced 120 miles apart. Each is equipped with five cameras that scan 250,000 square miles of prairie night sky. MORP is still waiting for its first meteorite recovery.

The Czech network was the first to recover a meteorite, the great Pribram, and to determine its orbit.

The Smithsonian Prairie Network in the United States, discontinued in 1975, also photographed one fireball — the Lost

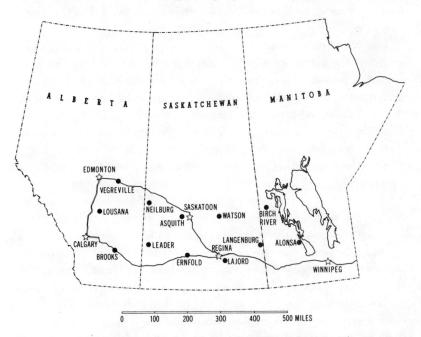

Figure 4. *The MORP network of meteor cameras operated by The National Research Council of Canada. (Courtesy National Research Council.)*

City — with such precision that the meteorite was found and its
path in space could be accurately known. The Lost City fell in
Oklahoma January 3, 1970, at 8:14 P.M. (C.S.T.) At its most
distant point, it reached 2.35 times the earth's distance from the
sun. The Lost City meteorite is the type known as a bronzite
chondrite, showing flow patterns in the fusion crust. Orbits of
fireballs show that they are material that has probably come from
the asteroid belt, between Mars and Jupiter, broken pieces of a
larger body.

Finds of meteorites are usually the sturdy nickel-iron variety
which can survive for centuries on the earth. In the case of the
largest ones, the recovery makes a saga in itself. The Willamette
meteorite is a 15½-ton mass of iron and nickel now resting in the
Hayden Planetarium in New York. In 1902 a Welshman, Ellis
Hughes, and a prospector named Dale were searching the
Oregon hills for minerals. They saw a point of rock projecting
from an area of level rock. A hammer blow showed that the point
was iron.

The two men thought they had found an iron mine, and
guarded the secret carefully. They tried in vain to buy the land
from the company that held it. Then they discovered that their
iron mine was only 10 feet long and about a yard deep, and they
realized they had found a meteorite. By August, 1903, Dale had
departed from this continent and Hughes conceived the idea of
transporting it to his home three-quarters of a mile away.

To help him in this endeavor he had one small horse and one
15-year old boy, his son. He made a wagon of log timbers
mounted upon cross-sections of tree trunks for wheels. And he
built a simple capstan, anchored it to the ground and attached a
100-foot wire rope, carefully braided. As the horse circled around
it, this would wind up on his primitive winch. Hughes managed
to overturn the great iron mass onto his cart and lash it there. For
three months the caravan pursued its weary way, straining and
pulling, toward the back yard of the Hughes home. On some
days when the wheels sunk in soft mud they made only several
feet; on the better days as much as a hundred feet.

But a meteorite of 15½ tons is a rather unusual back yard
ornament. People came from miles around to see it. A natural

consequence was that the Portland Land Company, on whose property the meteorite originally rested, brought suit to reclaim its treasure. Although public sympathy was rather with Hughes, he lost the case and the company regained the small iron mine. They had no particular use for it, and sold it to Mrs. W. E. Dodge, a New York philanthropist, for 20,600 dollars. Mrs. Dodge generously and sensibly presented it to the American Museum of Natural History, where millions of people can view the giant from the skies.

The Willamette is not the largest meteorite to be seen in a museum, however. That honor belongs to the Ahnighito or Tent meteorite, also on display in the American Museum, on a special scale showing its weight to be 68,085 pounds, or 34 tons. The largest meteorite known, the Hoba West, still lies unmoved and unweighed on the African veldt near Grootfontein.

Among the most thrilling stories of finding and recovering meteorites is that pertaining to the Ahnighito and its companions the Woman and the Dog, also known as the Saviksue, Great Irons or Cape York meteorites. The story begins back in 1818 when Captain John Ross in his arctic explorations found a previously unknown tribe of Eskimos near Cape York, Greenland, who were using knives and harpoon points with edges of iron. Despite communication difficulties with these Eskimos, Ross learned that the metal had come from an "Iron Mountain" on the north shore of Melville Bay. Analysis showed the presence of nickel, and a meteorite was inferred.

To solve the mystery of this Iron Mountain was one of the aims of every expedition into that region from 1818 on. No more information was obtained until May, 1894, when Captain Robert Peary gained the confidence of the Smith Sound Eskimos. One of them, Tellikotinah, guided him to the Iron Mountain. It turned out to be three large masses of homogeneous metal, which Peary could not carry away. The Eskimos described the Saviksue as an Inuit woman and her dog and tent hurled from the sky by the Evil Spirit, Tornarsuk. Such a legend indicates that the stones came down while the region was inhabited. Originally the Woman had the shape of a woman sitting at her sewing. Constant chipping of fragments, however, had removed the upper part of her body, reducing her to one half or one third her original size. The story is that a band of Eskimos from a settlement north of Whale Sound tried to carry the head home, to have a supply of the precious metal closer to them. But on their way home the sea ice suddenly broke up, and the sledge with the head lashed on disappeared beneath the water, carrying the dogs with it. The Eskimos managed to survive, but the incident discouraged them from any further attempt to carry away anything but small pieces of the "Star Stones."

As soon as Peary learned the location of the Star Stones, he made many persistent attempts to bring them back to civilization. His description of his efforts given in *Northward Over the Great Ice* is dramatic indeed. The meteorites were on the northern shore "of that great icy fastness, Melville Bay, some thirty-five miles east of Cape York . . . In winter this region is the desolation of Arctic desolations, constantly harassed by biting winds, and every rock deep buried beneath the snow, swept in by these winds throughout the long dark night, from the broad expanse of Melville Bay, and piled in drifts, which in many places are hundreds of feet deep."

In August of 1894 Peary attempted to reach Melville Bay in his ship the *Falcon*. But that summer the ice did not go out of the bay and he could not get within 30 or 40 miles of the meteorites. He decided to winter in the Arctic. In December he tried to reach them by dog sled, but in vain.

He returned in August, 1895, in the steamer *Kite*. When he first entered Saviksoah Bay the winter's ice was completely across it. Finally the ship was able to enter a small channel, and Peary went up the little valley and "stood once more beside the great heaven born mass."

The two smaller meteorites, the Woman and the Dog, lay loosely on the rocks of an eastern arm of the bay. Six miles to the south, the third and largest, the Ahnighito or Tent meteorite was nearly buried in the rocks and soil. It was 100 yards from the shore and some 80 feet above high-water mark. The surface of the meteorites, dark brown with greenish bits, looked like bronze. Later analysis showed a similar composition for all three, characteristic of the nickel-iron type of meteorite, about 91 percent iron, 8 percent nickel with traces of cobalt, copper, sulfur, phosphorus and carbon. Though the chemical analysis of the three shows only minor differences, there appears to have been a great deal of difference in their amiability for working. The Woman was surrounded by a great pile of broken trap cobble. Only a few stones were scattered about the Dog, and none about the Ahnighito, which seems not to have been worked at all.

The Dog and the Woman did not present too great a removal problem for Peary. The Dog, the smallest of the Saviksue, is about 27 by 19 inches, weighing 1,000 pounds. The next largest, the Woman, is four feet long, three feet wide and two feet thick. Its estimated weight was 6,000 pounds. After several days' work with the ship's crew and Eskimos, and the help of jacks and a sledge of spruce poles, the two meteorites were carried toward the ship. The Woman was placed at the shore on a huge cake of ice, seven feet thick, 40 feet long and 20 feet wide. This was used as a ferryboat across the open water, where a dock was cut for it in the bay ice. Peary recounts that all possible speed was made in hooking the ship's tackle and purchases on to the meteorite. Before the job was finished, the ice began to give way and tense moments prevailed until the "sulky giant" was finally landed in the hold.

At the same time Peary and his party began a partial excavation of the Tent in preparation for its removal, but the problem was too great for completion then. Undaunted, Peary secured a larger ship, the *Hope,* of 307 tons register, and went north again in July, 1896, with heavier equipment. He stopped long enough at Cape York to take on board all the able-bodied men of the village. Then he continued eastward across Cape York Bay to Saviksoah Bay and the eastern side of Meteorite Island. For 10 days, through night and day, the band worked on the meteorite. The great jacks tore it from its bed, and by ropes and timbers it was sent down toward the shore. Peary comments that the meaning of the terms "momentum" and "inertia" had never been brought home to him so forcibly as in handling the mountain of iron. And it seemed to have a touch of the supernatural about it. On the last night of the stay, when a driving snow came down in horizontal lines, everything around was buried in the falling snow except the Ahnighito, which towered above the human figures silhouetted around it. The flakes of snow vanished as they touched the Saviksoah, as though the meteorite still held some of its celestial fire. But the insweeping ice forced Peary once more to abandon his prize, and to hope that he could return for it.

The opportunity came the following year when he again reached Cape York in the *Hope*, on August 12. The ice conditions were good, and he immediately took the ship to the berth alongside the Saviksoah on Meteorite Island. With the help of Captain Robert Bartlett he was able to maneuver the ship to within about 18 feet of the shore alongside the Ahnighito. Then the workers fashioned a bridge of two 14-by-16-inch timbers of white oak, 60 feet long, to span the gap between the ship and the shore. They reached under the meteorite at one end and across the ship at the other. After five days and nights of labor, at 6 P.M., Friday, August 20, 1897, the meteorite was safely placed above the ship's waist.

When the tide was exactly right the meteorite, draped in Old Glory, was started down toward the ship, and Peary's little daughter christened it Ahnighito with a bottle of wine. Many moments of anxiety remained until the iron monster was securely settled in the bowels of the ship and in such a posture that its great weight would not topple the vessel as she buffeted her way through the ice. The efforts of years of hardship and waiting came to an end on Saturday, October 2, 1897, when the ship reached the New York Navy Yard. There the hundred-ton floating crane lifted the meteorite out from the *Hope*.

Actually the recovery of all the Saviksue was not completed then. In 1913 a fourth named Savik, weighing eight tons, was found 15 miles from the others. This was taken to Copenhagen in 1925.

There are many fascinating tales of meteorite recovery. Two of the unhappy ones carry warnings of what *not* to do if you are lucky enough to find a meteorite. Near Port Orford, Oregon, in 1859 a sample from a 10-ton rock on a mountainside showed typical stony-iron meteorite structure. Soon after the discovery the only person who knew the location died. To this day the meteorite has not been refound. Another unhappy story comes from California. Some time before 1887 a prospector found an 80-pound stony meteorite in the San Emigdio Mountains. Thinking it was an ore of precious minerals, he sent it to a geologist, who put it through an ore crusher and pulverized it. Only three or four tiny pieces of it, totaling five ounces, exist today.

A meteoric event on a colossal scale, and of a unique nature, occurred in Siberia early in this century. This is the Tunguska event, also known as the great Siberian Cometoid. The flash of light could be seen in a cloudless sky in central Siberia over an area about 1,000 miles in diameter. The sound phenomena were heard up to 800 miles from the explosion center. The resulting devastation was far greater than any other caused by a celestial visitor in historic times. Fortunately this happened in a wilderness area.

On June 30, 1908, about eight o'clock in the morning, local time, the inhabitants of the region 60 miles northwest of Vanovara were startled by a blinding flash, comparable to a conflagration, followed by darkness and an explosion. A farmer, S. B. Semenov, who was sitting on his open porch facing the north, had his shirt almost burned off with the great heat. The explosion blew Semenov seven feet off the porch, leaving him unconscious for a brief while. When he came to, he was surrounded by crashing stones, and the houses were shaking and moving from their foundations.

Despite the magnitude of this event, because of the desolate, nomadic character of the region and the occurrence of World War I, many years elapsed before it was scientifically investigated, first by L. Kulik in February, 1927. Penetration of the country is difficult, with swampy forests (taiga) overlying permafrost in summer, and severe arctic conditions in winter. The center of the explosion (ground zero) was at latitude 60°54' N., longitude 101°57' E., in the basin of the upper part of the Podkamennaya Tunguska river, 400 miles north of the Trans-Siberian railroad, between Venissei and Lena.

Within a radius of 20 miles of ground zero, virtually no trees were left standing, and as far as 37 miles from the center of the explosion broken trees were observed. An estimated 180 million trees were felled, and the region was strewn with freshly formed craterlets.

All the former vegetation shows the effect of a uniform, continuous burning. It was little short of miraculous that apparently no human lives were lost in this catastrophe. A herd of 1,500 reindeer belonging to a wealthy Tungus, Vasily Ilyich, were not so fortunate, however. These reindeer roamed through the valley of the river Khushmo. After the catastrophe, when the Tunguses went hunting for them, they found a few charred carcasses, but of most of the reindeer — nothing. The sheds, too, had burned and melted and only a few buckets remained.

Astronomers, both Soviet and others, agree that this must have been a cometary body of ice and dust, and not a rocky meteorite, that came toward the earth. Its mass is estimated at 200 tons. Vaporized and heated to high temperatures, this could set up the effect of a hurricane with intense heat. Changes in atmospheric pressure were registered on barographs all over the world. Some, timed at the Potsdam Geodesic Institute, indicated that the body had its main explosion at a height of three miles up in the atmosphere. The ingenious hypothesis that a "black hole," with a mass of many tons, but less than a millionth of an inch in diameter, was responsible, instead of a comet, does not explain some of the observed facts.

Over western Europe on June 30 and July 1 of that year a dust cloud brightened the sky from 50 to 100 times its normal night brightness, to the extent that farmers could gather crops. Significantly, on June 29 no such abnormal brightening was seen, and by July 2 it had vanished. This dust is considered to be the tail that accompanied this tiny comet.

The products of the explosion produced another kind of dust, which spread gradually over the entire northern hemisphere. It actually diminished the amount of solar radiation received in the northern hemisphere. The dust reached California two weeks later. The gases released by the explosion would be those common to comets and the atmosphere — carbon dioxide, water, methane and ammonia. There might be small amounts of the deadlier gases in comets, such as cyanogen, but too little of these to have any lethal effect.

The chances that such a devastating celestial happening will occur over heavily populated areas, or over some of the world's great cities, are very, very small — but they are not zero. It is important that, in moments of panic, people realize that unpredictable devastation can be caused by nature itself, and not attribute it to a national enemy.

The Great Meteoric Procession of 1913

One other meteoric event within living memory is also a unique type. On February 9, 1913, residents of south-central Canada were treated to a celestial display the like of which had never been recorded before, and has not been seen again.

Around Toronto the display came just after nine o'clock that winter Sunday evening. As reports poured in by telephone to the newspapers (but not to radio stations!), Dr. Clarence A. Chant, head of the Department of Astronomy at the University of Toronto, realized that a most unusual event had occurred. With the cooperation of the press he managed to gather as many fresh eyewitness reports as possible and weave them into a coordinated story.

At 9:05 P.M. that night, as seen around southern Ontario, a fiery red body appeared in the northwestern sky. As it came nearer it grew larger and was followed by a long tail, described as fiery red or golden yellow. Some said the tail was like the glare from the open door of a furnace. Others said it was like a searchlight or a stream of sparks blown away from a burning chimney by a strong wind. Many people thought someone had set off a gigantic skyrocket. But the body differed from a rocket in that it showed no inclination to drop to earth. Instead, it moved forward on a horizontal path with peculiar slowness. Without any apparent sinking toward earth, it disappeared in the distance in the southwest.

Before observers had recovered from their astonishment, other bodies appeared from the northwest in precisely the same place.

They came on at the same slow pace, in twos, threes, or fours, with tails streaming behind them, but not quite so bright as the first, and all headed for the same point in the southwestern sky. Gradually the bodies became smaller. The last ones seemed to be just red sparks that were extinguished before they reached the endpoint of the path. As the bodies were on the point of vanishing, a rumbling sound was heard in many places. It was likened to distant thunder or a carriage passing over rough roads. A noted Canadian artist, Gustav Hahn, was fortunate in seeing the display. The David Dunlap Observatory in Richmond Hill, Ontario, is equally fortunate in having his painting — of high standard, both artistic and scientific — showing the meteors heading toward the beautiful winter constellation of Orion.

The outstanding feature of the procession was the slow, majestic motion of the bodies and the perfect formation in which they moved. Some compared them to a fleet of airships, with light on both sides and forward and aft. Others saw a similarity with battleships, attended by cruisers and destroyers.

Estimates of the actual number of bodies ranged from 15 to 20 up to 60 or 100. Some said as high as thousands. Only one person had an opera glass in hand, Master Cecil Carley, a high school pupil. He said there were ten groups in all, and as seen through the opera glass each group had 20 to 40 meteors. The entire time for the procession to pass one observer was about 3.3 minutes.

When more reports came in, Dr. Chant was able to show that the procession had been seen from Saskatchewan in the west, across southern Canada, to Bermuda in the southeast, covering a distance of 2,500 miles. From such a long track, as well as the slow motion, Chant drew the conclusion that these bodies, traveling through space, had come near enough to the earth to be captured by it. They were actually moving about the earth in a satellite orbit. These were, then, apart from the moon, the first natural observed satellites of the earth.

There was one puzzling point about this interpretation that provided an argument against it. If these bodies were really moving in satellite orbits, then a person would expect to see them rise up from the northwestern horizon, cross the sky and go down again to the horizon in the southwest. But this was not what observers saw. The bodies first appeared at altitudes well up from the horizon. Almost half a century later this puzzle was solved. When the first artificial satellites, the sputniks, were placed in the sky, observers saw them behave in just this way. They were not seen to rise from the horizon and then sink right down back to the horizon. On the contrary, they seemed to come at you out of thin air, at altitudes of 10° or more. Atmospheric thickness at lower altitudes plays a part in their appearance and disappearance.

This finding naturally kindled interest again in our natural earth satellites. Dr. John O'Keefe of the Goddard Space Flight Center made a study of the procession. He agrees with Chant's earlier finding that the bodies fell into the Atlantic Ocean off Bermuda or even off the coast of South America.

Chant estimated the diameters of the larger bodies to be as much as 100 feet. They were traveling with respect to the earth's surface at a speed greater than five but less than 10 miles a second. What would these bodies have looked like had they landed on solid ground, not ocean? No one knows. Is it possible that the material coming from such a procession would resemble those mysterious glassy disks, tektites, whose source is still unknown?

Tektites remain one of the greatest puzzles among rocks found on earth. Where and how did they originate? The word "tektite" is derived from the Greek *tektos,* meaning melted. Tektites are small bodies of silicate glass found in a few regions of the earth's surface known as "strewn fields." Flow marks are an important characteristic, presumably obtained in their passage through the earth's atmosphere. Tens of thousands of them have been found. Analysis shows them to have properties that distinguish them from any other rocks, though in superficial appearance many resemble natural obsidian glass.

Their distinctive shapes, some resembling buttons, arise from their origin as spherical drops. Sometimes their spin produces characteristic forms of oblate spheroid or dumbbell. Aerodynamical friction also can shape them, as well as terrestrial erosion as they lie on the earth for millions of years.

Tektites from a given strewn field have a family resemblance to each other, but have different characteristics from those of other fields. The best known are the moldavites from Czechoslovakia, the australites from Australia and Tasmania, the indochinites from Indo-China, and, in lesser quantities, the bediasites from Texas.

The theory that the tektites are splashes of lunar material was not confirmed by lunar exploration. Not yet ruled out are theories which suggest they were formed when a comet's head or a giant meteor approached close to earth.

Scars on the Earth's Surface

The planets and moons in the solar system which have visible, rocky surfaces bear many scars of bodies that came hurtling at them from space. The earth, Mars, Mercury, Venus, our moon and the moons of Mars are all scarred. Of these bodies, our moon bears the most scars, hundreds of thousands of them. Splashes from one hit can make many craters. There is some evidence that such impacts occurred more frequently in the first billion years of the solar system, something for which we have no regrets.

The earth's atmosphere is dense enough to save it from many hits by smaller rocks, but it will not prevent bodies of many tons coming down. However, the weathering effects of our air blanket and geological processes do erase these scars over long periods of time. Even so, the earth seems to have fewer visible impact craters than one would expect. The Canadian terrain is particularly favorable for their preservation, and Canada has been in the forefront of countries discovering such craters.

Impact craters fall broadly into two types. The first is the crater with sizable pieces of meteoritic material in or near it. That is accepted by everyone as a hit from outer space. The other type is a crater that shows the type of shock formation that could have been formed only under the high pressures of impact, not from geological processes on earth; however, the most convincing proof — that of meteorites themselves — is lacking. All of these craters are prehistoric, and in fact most of them were formed hundreds of millions of years ago. They are often called fossil meteorite craters. No one disputes that any meteoritic material would have disintegrated in the intervening eons.

The first of these great scars to be recognized for its true origin was the Barringer Crater in Coconino County, near Winslow, Arizona. This is also the best known, at least in North America. The crater, formed in solid limestone and sandstone, is four-fifths of a mile in diameter, 600 feet deep, with the rim rising 125 feet above the level of the surrounding plateau. It is not conspicuous from the surrounding countryside toward which its outer walls gently slope, but is much more noticeable from the air above. Early in this century Daniel Moreau Barringer, a geologist and mining engineer from Philadelphia, proposed a meteoric origin for it. Meteoritic fragments were found on all sides up to distances of four or five miles from the crater, but more numerous closer to it. They showed half an ounce of platinoid materials per ton. These are often called the Canyon Diablo meteorites from the stream a short distance away.

In early drilling exploration near the start of this century, the first hopes were that the main mass of the meteorite, presumably still buried in the center of the crater, might provide a ready-made mine, perhaps for platinum. No huge mass has ever been located and it is unlikely that one exists. Over 20 tons of meteoritic materials have been picked up. The estimates of the size and mass that hit the earth depend upon the assumed velocity. Eugene Shoemaker of the U.S. Geological Survey suggests that, with a probable velocity of 10 miles a second, the mass could be 63,000 tons, the diameter about 81 feet. The energy of the explosion would be equivalent to 1.7 megatons of TNT. The hit occurred some tens of thousands of years ago. Even though the crater is circular, the massive body came in at a low angle from the north.

Less impressive craters from smaller falls are found in many parts of the earth. Near Odessa, Texas, a crater 500 feet in diameter and 16 feet deep is associated with nickel-iron meteorites. At Campo del Cielo, in Argentina, there are several craters, the largest with a diameter of 250 feet, and several tons of meteorites have been recovered. In northern Chile, from a study of air photos, a large crater was found recently by Joaquin

Sanchez and Dr. William Cassidy of Columbia. It is 1,476 feet in diameter, with an average depth of 100 feet, in a foothill area 10,000 feet above sea level.

Near Wabar, in the Arabian desert, are two craters 325 feet and 150 feet in diameter, with nickel-iron meteorites around. Some famous craters, six in all, with the largest 325 feet across, are on the island of Saarema (Oesel) in Estonia, with meteoritic iron around. In Australia there are four well-established meteorite hits with 17 individual craters, at Henbury, Boxhole, Dalgaranga and Wolf Creek. Of these the Henbury are the best known with 13 craters, ranging from 600 to 30 feet in diameter, with meteorite fragments in some.

The extensive and successful search for meteorite craters in Canada followed the discovery of a large crater in Ungava by Fred Chubb, a Canadian prospector. In February, 1950, he showed Dr. V. B. Meen of the Royal Ontario Museum a set of aerial photographs taken by the R.C.A.F. in 1946, in the northwestern tip of the province of Quebec, a region known as Ungava. On these photos was a crater-like lake 10,000 feet in diameter with a rim height of 500 feet. The rim ruled out the possibility that the lake could be interpreted as a sinkhole. Two other origins were possible: an extinct volcano or a meteorite crater. The volcano might prove to be a diamond pipe, the missing source for diamonds that are found over the eastern United States in glacial gravels coming from northeastern Canada.

The possibility of such a discovery meant that the first expedition, headed by Meen and Chubb and financed through K. W. MacTaggart of the Toronto *Globe and Mail*, had to be shrouded in secrecy. On July 21, 1950, the expedition reached the rim of the crater. Snow still remained on the north slopes of the crater rim, and three-quarters of the surface of Crater Lake was covered with floating ice three feet thick. The rim rises out of the lake at an angle of approximately 45° in most places. It is a mass of granite bed-rock heaved to its present location by an enormous explosion. Other expeditions have since gone into the area and have confirmed the meteoritic origin of this impressive feature, named the New Quebec Crater, but no meteorites have ever been recovered. In 1953 J. M. Harrison determined that the

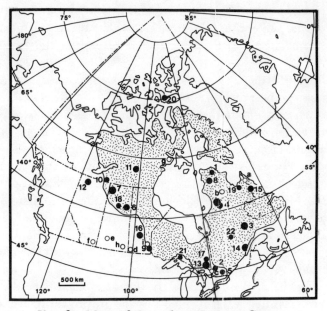

Key for Map of Canadian Impact Craters

Confirmed Sites

1 New Quebec Crater, P.Q.
2 Brent, Ont.
3 Manicouagan, P.Q.
4 Clearwater Lake West, P.Q.
4 Clearwater Lake East, P.Q.
5 Holleford, Ont.
6 Deep Bay, Sask.
7 Carswell, Sask.
8 Lac Couture, P.Q.
9 West Hawk Lake, Man.
10 Pilot Lake, Mackenzie
    District, N.W.T.
11 Nicholson Lake, Keewatin
    District, N.W.T.

12 Steen River, Alta.
13 Sudbury, Ont.
14 Charlevoix, P.Q.
15 Lake Mistastin, Labr.
16 Lake St. Martin, Man.
17 Lake Wanapitei, Ont.
18 Gow Lake, Sask.
19 Lac La Moinerie, P.Q.
20 Haughton Dome, Devon Island,
    N.W.T.
21 Slate Islands, L. Superior
22 Ile Rouleau, P.Q.

Probable Sites

a Merewether Crater, Labr.
b Kakiattukallak Lake, P.Q.
c Skeleton Lake, Ont.
d Hartney, Man.
e Elbow, Sask.

f Eagle Butte, Alta.
g Meen Lake, Keewatin, N.W.T.
h Viewfield, Sask.
i Poplar Bay (Lac Du Bonnet).
    Man.

Figure 5. *Location of 23 confirmed and 9 suspected Canadian fossil meteorite craters. (Courtesy Earth Physics Branch, Department of Energy, Mines and Resources.)*

crater had undergone glaciation, and this would render fruitless a search for meteorites. Harrison notes that the crater must have been formed before the last advance of continental ice in this area.

Following the discovery of this crater, scientists at the Dominion Observatory, Ottawa, under the direction of the Dominion Astronomer, Dr. C. S. Beals, began a study of 3 million aerial photographs of Canada. A total of more than 40 formations, circular or partly circular, were identified as worthy of investigation. Great care was taken to discard normal features and to select only those with raised rims or other variations from normal. Many of those selected are on the Canadian Shield, which has an area of 1,800,000 square miles or about half the area of Canada. The Shield is composed of Precambrian rocks. The youngest of these are at least 500 million years old, but most are much older, from 1.5 to 2 billion years.

To date, 23 of these craters have been sufficiently studied that they are classified as confirmed impact sites by P. B. Robertson and R. A. F. Grieve of the Earth Physics Branch, Department of Energy, Mines and Resources, Ottawa. These are shown on the map on page 51 and listed in the Key, along with nine probable sites, as yet unconfirmed. The confirmed impact craters all give evidence of shock metamorphism, produced only by high-velocity impact. Most of these hits occurred hundreds of millions of years ago. Many of the craters are much larger than the meteorite craters described earlier, and approach in size some of the well-known lunar craters.

The best known and one of the largest is the Sudbury basin. As the largest known nickel deposit, this is the source of most of the world's nickel. Dr. Robert S. Dietz of the U.S. Navy Electronics Lab at San Diego first proposed in 1964 that this was caused by the hit of a small asteroid. He called it an astrobleme, a star-wound. The strike caused a shock that triggered the rising of molten lava from the earth. The explosion, with the force of 3 million megatons of TNT, blew out a crater into which the molten rock poured. Dr. Dietz has remarked whimsically that pennies may not come from heaven, but nickels do.

The hypothesis was slow to be accepted because there are

many evidences for ordinary geological processes in the Sudbury basin. Now it seems that the riddle is solved by a determination of an age of 1,800 million years for the impact structure with complex geological changes occurring after that. The original crater may have been as large as 60 miles across, bigger than the Sudbury basin.

One of the smaller fossil craters is easily reached by car. The Holleford crater is about 15 miles north of the city of Kingston on Lake Ontario. Holleford is located a few miles east of Highway 38, north of Hartington and south of Verona, just west of the northern tip of Knowlton Lake. The small crossroads of Holleford is in the center of the crater, and as you drive through, you can see the remains of the crater walls. The mineral coesite, a proof of a great collision, has been identified in this crater. It was first found in a very old feature in Central Europe, the Ries Kassel, giving proof that this also was a meteorite hit.

A Canadian crater that is easy to see from the air is White Hawk Lake, in southeastern Manitoba, because the transcontinental planes frequently go over it.

A very noticeable feature on the map of Canada is the great circular arc forming the eastern side of Hudson Bay. Known as the Nastapoka Island arc, this feature is unique on the surface of the earth, according to Dr. C. S. Beals, who has investigated it. Beals sees it as one edge of a great meteorite crater, comparable in size with some of the lunar seas. Geophysicist Dr. J. Tuzo Wilson relates it to natural geological processes, which might produce such a formation. Drilling and other modes of investigation of this arc are exceedingly expensive and time-consuming, so it may be a long while before the origin of the Hudson Bay arc is determined.

★

★

★

THREE

The Moon

Soon as the evening shades prevail the moon takes up the wondrous tale.
And nightly to the listening earth repeats the story of her birth.

ADDISON, Hymn of the Creation

The very word "moon" brings to mind those moments when our earth is made particularly beautiful by moonlight. You may recall a night before Christmas when "the moon on the breast of the new fallen snow gave the lustre of mid-day to objects below." Or you may think of a silver sliver of a crescent moon near the horizon, or a huge golden moon rising in the east. Perhaps the fact that the moon comes and goes in our skies, while the sun shines steadily on day after day, gives the moon extra glamor. Or perhaps its dimmer light, which permits us to look directly at it, adds to its charm.

Phases of the Moon

The presence or absence of the moon in our skies is directly tied to its phases, which are noticed by practically everyone. There is evidence that primitive man was conscious of the phases of the moon tens of thousands of years ago. Science writer Alexander Marschack has so interpreted notations and markings extending backward from the Mesolithic Azilian to the Magdalenian and Aurignacian cultures in Europe, as much as 35,000 years before history.

The actual explanation for the phases of the moon is less well known than the fact that they exist. For every phase of the moon there is a rigid relation between the positions of the earth, sun and moon. Artists and writers take innumerable liberties — indeed, too much poetic license — in putting the moon at some phase like crescent or full into a position it cannot possibly occupy in the sky. In fact, an article in the July, 1975, issue of a popular magazine places the waning moon in the early evening western sky, setting before midnight. Only the waxing moon occupies this position.

The moon has no light of its own, so the light it sends us is all reflected sunlight. It acts like a giant mirror, and if it happens to be turned so that the sunlight falling on it is not sent entirely in the earth's direction, we see it as other than round. The moon is a

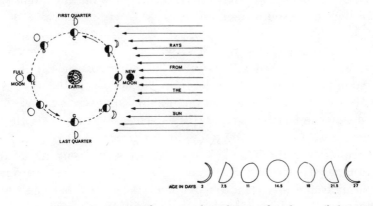

Figure 6. *Relative positions of sun, earth and moon for phases of the moon. Lower diagram shows age at given phase.*

very poor reflector, however, sending off only 7 percent of the light falling on it. (If we had such a decrepit mirror around the house we'd either have it resilvered or throw it away.)

Obviously 50 percent of the moon's surface will be in direct sunlight at any time, just as 50 percent of the earth's surface is. How much of the bright 50 percent of the moon is turned toward earth determines the phase.

When the moon is lined up between the earth and sun, the sun is shining on the rear side of the moon, the side we cannot see, and the phase is new. (Sometimes the moon does get in direct line with the sun and eclipse it.) The yearly almanacs give the time, usually expressed to the nearest minute, when the moon will be new. The interval that has elapsed since new moon is called the moon's age. Right at new moon this is zero days.

After the new phase, when the moon has passed conjunction with the sun, it continues its eastward motion among the stars, becoming visible later and later in the evening sky. You have noticed the very young crescent moon in the west right after sundown. That moon has been in the sky all day, rising shortly after sunrise, but you did not notice it until the glare of the sun was extinguished. As the moon moves farther east, you will notice it higher in the sky and farther east after sunset. The crescent is waxing.

About seven days after new moon, the crescent has become a half-moon, or first quarter. This is the phase of the moon that you see almost due south when the sun sets. It sets in the west around midnight.

When another seven days have passed, during which the moon has become gibbous and moved still farther east, it will be full. The full moon will rise in the east as the sun sets in the west, and will be above the horizon all the time that the sun is below. The full moon and the sun are mutually exclusive in the sky. With the moon now waning all the time, in seven more days the third or last quarter is reached. This time we see a different half of the illuminated portion. The last quarter moon rises around midnight and sets around noon. You may see it as a large white moon in the western sky when you go about your business in the morning.

The waning moon then shrinks down to the very narrow crescent of the very old moon, with an age of 26 or 27 days. This is the crescent that you notice in the east before sunrise. The very old moon rises an hour or two before the sun, crosses the heavens in front of it, and sets in the west ahead of it.

The moon goes around the earth in an average time of 27 1/3 days, its sidereal period. During this time, however, the earth has been moving too. So it takes a couple more days before the moon will line up again with the sun. This is the interval of the moon's phases, 29½ days, known as its synodic period. So every 29½ days, on an average, the phases of the moon repeat themselves.

Now you can see why the moon's position in the sky at a given phase is very limited. Romantic as it might sound for a novelist to have the full moon rising over the eastern horizon as the clock strikes midnight, it just cannot happen. The full moon has to rise approximately at sunset. It cannot be rising above the horizon at other times. (In the High Arctic or Antarctic you might get some queer moon-rising and setting intervals, just as with the sun.)

When you see the young crescent moon in the west just after sunset, you can also see the dark part of the moon shining with a misty glow. This beautiful phenomenon is called "the old moon in the new moon's arms." It is caused by sunlight reflected from the earth to the parts of the moon not in direct sunlight. The moon and the earth show opposite phases to one another; that is, when the moon looks new to us, the earth will be full as seen from the moon. Because of its larger size, the earth sends the moon much more light than the earth ever receives from it. The first person to explain in print that there is earthshine on the moon was the versatile Leonardo da Vinci.

Motions of the Moon in the Sky

The motions of the moon and the sun, and in fact of all the bright celestial objects, make us conscious of the daily turning of the earth on its axis. For all of these bodies appear to rise in the east, cross the heavens and set in the west for the simple reason that we are being whirled on the surface of the earth toward the east by

The path that the moon follows in its circuit of the heavens is very similar to that of the sun, described in the next chapter. The sun's apparent path is called the ecliptic. The moon's path is similar to the ecliptic, but inclined 5° to it.

During one month the moon takes up the positions in the sky that the sun assumes during one year. Therefore, just as the sun rides high in the sky at one interval each year, in summer, so some phase of the moon will ride high each month. Just as the sun runs low in the sky in winter, so each month some phase of the moon will run low. Since the new moon is close to the sun in the sky, it will ride high in summer and low in winter. Since the full moon is opposite the sun in the sky, it will ride high in winter and low in summer. You have probably noticed when you are outdoors in summer, maybe at cottages or camps, that the full moon never seems to get very far up in the sky — it coasts along the treetops of the southern horizon, pursuing the same path that the sun has in midwinter.

Though there is not much difference between the moon's path and the ecliptic, that 5° inclination does have several important effects. It prevents an eclipse always occurring at every new or the earth's turning on its axis. This means that all the heavenly bodies appear to rise in the east and set in the west once in approximately 24 hours. Astronomers speak of this as diurnal motion.

But some of these bodies have apparent motions of their own that are large enough to be noticed, and these motions are in a different direction. The moon moves eastward approximately by its own apparent diameter, one-half a degree an hour, making a complete 360° circuit of the heavens in 27 days. Because this eastward motion of the moon is small compared with the broad sweep of diurnal motion to the west, it is masked by the diurnal motion unless you look closely. On a night when the moon is near a bright star or planet, note its position and look at it again several hours later. You will notice that the heavens have turned toward the west, carrying the moon along with them. While this was happening, the moon has been slowly crawling across the sky in the opposite direction, toward the east. It will now be in a different position with respect to the bright star or planet.

full moon. It gives the moon a greater swing in meridian altitude than the sun has, and also in its azimuths of rising and setting. (The azimuth is a measure of the number of degrees from the north point (0°) along the horizon eastward.) On certain years the moon can be as much as 28½° either north or south of the celestial equator. This brings it 5° nearer the zenith (the overhead point) than the sun ever gets in the temperate and frigid zones. And the converse is true also. The lowest altitude that the moon has when it crosses the meridian can be 5° less than the lowest meridian altitude of the sun in that particular place. In other words, if you think of the position of the sun at noon on a brief December day, the moon, even on the meridian, can be 5° (10 times its own apparent diameter) lower in the sky at certain times.

These high and low altitudes of the moon go together in cycles of approximately 19 years. They depend on the position of the points of intersection of the moon's path and the ecliptic called the nodes. We passed a period when the effect was most marked in the winter of 1968-69. By 1977 the nodes will be in such a position that they make the moon hug the celestial equator more closely than the ecliptic does. Neolithic man knew about this 19-year-cycle. Some stones at Castle Rigg near Keswick in the Lake District of England are oriented for the extreme northerly and southerly positions of moonrise.

The angle of inclination of the moon's path to the horizon varies greatly at different times and seasons. This variation has some noticeable effects. One of these is the harvest moon. Some persons think, erroneously, that the term "harvest moon" applies to a particularly large, coppery moon ("big as a dinner plate"!) on the eastern horizon, perhaps with a field of corn stacks and orange pumpkins silhouetted against it. The term comes from the fact that this moon gives farmers added hours of light in which to gather crops.

By definition, the harvest moon is the full moon that occurs nearest the autumnal equinox (either before or after). This moon near the full gives us more light in the early evening after sunset than any other, because it has a smaller delay in rising from night to night. On average the moon rises about 50 minutes later from

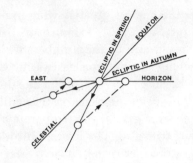

Figure 7. *Angle of ecliptic with horizon, responsible for small delay in moonrise for Harvest Moon.*

one night to the next. This 50-minute delay in rising is caused by the fact that the moon is 13° farther east from one night to the next. The earth therefore has to turn through about 50 extra minutes (on top of its 24 hours) before it can make the moon rise again.

The actual delay in rising is dependent on the tilt of the moon's path with respect to the horizon. There is a big variation in this angle for different phases of the moon every month. It turns out that, in the northern hemisphere, for the full moon that occurs nearest the autumnal equinox, the tilt is near a minimum. The earth then has to turn through much less than the average 50 extra minutes to get the moon above the horizon. In the latitude of New York it may be as little as 20 minutes. This means then that we have a succession of evenings of nearly full moonlight when the moon is above the horizon virtually all evening. In higher latitudes the variation in lag of moonrise is even more pronounced. At latitude 60° (near Churchill, Manitoba, site of extensive rocket programs) the moon may actually rise two minutes *earlier* from one night to the next, because it is close to becoming circumpolar for one day. The phenomenon of small retardation is repeated, though to a less pronounced degree, in the following full moon, which comes usually in October and is known as the hunter's moon.

By contrast, in the spring the moon near the full is hardly in the evening sky at all. The reverse effect is in operation. The angle of inclination of the path of the full moon to the eastern horizon is at a maximum, and so is the delay in rising. The delay may run as much as 80 minutes in the latitude of New York to practically two hours at latitude 60°. The moon near the full seems almost to whiz through the evening sky in March or April.

Some weather proverbs result from the varying inclination of the moon's path to the horizon. It is a curious fact that proverbs that have been handed down for centuries frequently *do* have some basis in fact. Have you heard of a "wet moon" or a "dry moon"? This usually refers to the tilt of the crescent moon's horns with respect to the horizon. Now this tilt depends upon the seasons. The line joining the horns, or cusps, is perpendicular to the direction to the sun (from the simple geometry of the phases of the moon). The line indicating the direction to the sun makes a high angle with the horizon after sunset in the late spring, and a low angle in the late autumn. At different places, these seasons are associated with different conditions. Some people think the moon is wet when the horns hold the water; others consider it wet when the horns are tilted to pour the water out. It all depends on where you live. You can choose your own system, but at a given season and for a given place the horns will have a similar angle with respect to the horizon. Proverbs connecting the moon and weather may have some justification, but in a less straightforward manner than the prophets themselves intended.

Other proverbs are connected with phases. "Plant your corn in the dark of the moon" is heard frequently in rural areas. Scientists tend to look on it with some amusement, but some farmers count heavily on it. After all, reason the scientists, how can the feeble little moon, 1/81 the mass of the earth and a quarter of a million miles away, influence crops? The answer is still being sought. Bird migrations are affected by the phase of the moon. Perhaps the farmer who plants his corn in the dark of the moon is getting it in one jump ahead of the crows who are not looking over his shoulder.

If the moon looks particularly large to you when it is near the horizon, you are enjoying "the moon illusion." By simple geometry the moon is slightly farther away when it is on the horizon than when it is high up in the sky. Therefore it ought to look smaller on the horizon and larger when high up or nearly overhead. Why doesn't it?

This apparent contradiction has been noticed for 2,000 years. Ptolemy of Alexandria tried to explain it in the 2nd century A.D. In the last two decades there has been an increasing interest in experiments to understand it.

The explanation does *not* lie in the effects of the earth's atmosphere. Refraction may alter the apparent shape of the moon, but it does not perceptibly increase its apparent size. Rather the illusion depends upon the combination of the human eye and the human brain. An object seen across earth terrain or a seascape looks larger than that same object seen across empty space, such as the moon near the zenith. This can easily be proved by photographs, on which the diameter of the moon measures the same, whether the picture is taken of the moon on the horizon or at the zenith. Also, if you close one eye and look at the moon through a cardboard tube with the other, it will look smaller at the horizon than without the tube.

The Tides

One thing that is no illusion is the influence the moon has on the earth in forming the tides. The tide-raising forces may be computed and predicted by mathematical formulae. Prehistoric man was probably aware that there is a connection between the moon and the tides because any astute observer at a seaside port can see, with simple timing, that there is a pattern between the moon's meridian passage and the time of high tide. On the average, each of these comes about 50 minutes later from one day to the next. This 50 minutes was dubbed, a long time ago, "the moon's earmark on the tides."

High tide does not usually occur at the time of the highest altitude of the moon. Various geographical factors influence the interval between transit of the moon and high tide. Actual observation is the easiest way of determining this interval, which is known as "the establishment of the port."

Of course, in one period of slightly over 24 hours there are two high tides, about 12 hours 25 minutes apart. The waters on the side of the earth toward the moon bulge toward it, and those on the opposite side bulge away from it.

The moon is not solely responsible for the tides, however. The sun has a tide-raising effect also, but a smaller one, because the tide-raising force falls in proportion to the cube of the distance of a body. So the moon, which is very near to us, even though it is of small mass, has more than twice the tide-raising effect that the massive sun has at its great distance. When the tide-raising forces of the two add together, as they do at new and full moon, there is a large range in the height of tides. These big tides are known as spring tides. You may have noticed that disastrous tides usually come near new or full moon. When, near quarter moon, the pull of the moon is somewhat offset by that of the sun pulling at right angles, the tides with a small range are known as neap tides. The tide-raising force of the moon itself varies, both with the altitude of the moon's meridian passage and with its position in its orbit. When the moon is nearest the earth at perigee, with a distance from the center of the earth that may be as small as 221,463 miles, its tide-raising force is about 20 percent greater than when it is farthest from the earth at apogee, when its distance may be as great as 252,710 miles.

One of the great sights of nature is the tidal bore thrusting into a shallow or constrained arm of the sea. The Bay of Fundy in Nova Scotia is one of the most remarkable of these. At times the bore has been a solid wall of water 50 feet high, pushing through the narrow channel.

The tides have a long-range effect on the earth. Observations of eclipses of two millennia ago, when compared with those made today, show that eclipses are now occurring about 3¼ hours too early. The day is lengthening because the rotation of the earth has slowed down. It seems like a miniscule amount — only 0.0016 seconds a century. But in the course of time it could have a major effect on the earth-moon system, even to the point where the earth is forced always to turn the same side toward the moon. Tidal friction in the shallow seas such as the Bering Sea is one of the major factors here. In eons past, the tidal pull manipulated the moon into keeping the same side toward the earth.

The Face of the Moon

For centuries man with his unaided eye has seen markings on the surface of the moon, the familiar "man in the moon." These dark markings may be recognized also as a crab or a girl reading a book. The Haida Indians of Canada's west coast see them as a moon god. Whatever resemblance they may have for you, they are permanent features of the moon's face. Sometimes people don't realize that these markings are always the same from night to night. This may be because their orientation in the sky changes as the moon rises in the east, crosses the heavens and sets in the west. Their positions relative to one another do not change.

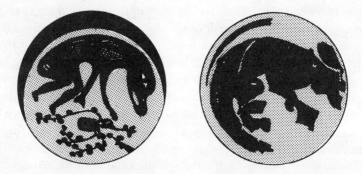

Figure 8. *Different people see different shapes for the man in the moon. Left: Moon god of Haida Indians, B.C. Right: Girl reading a book.*

The dark markings are the maria, or seas, a term which is a classic misnomer from the 17th century because there is no water on the surface of the moon. There are more than a dozen maria, and most of them are connected. The largest is dignified with the name of ocean, Oceanus Procellarum, Ocean of Storms, a landing site for some lunar probes. A very neat mare is Mare Crisium, Sea of Dangers, a circular sea by itself on the eastern edge of the moon. Probably the best known and most often discussed is Mare Imbrium, Sea of Rains, where a giant meteorite or small asteroid hit in the early years of the moon's existence. Lava then rose up and left its permanent marks on the moon's face.

Another feature you can see with the naked eye is the jagged edge of the terminator, the line between the sunlit and dark side of the moon. The roughness of the edge is caused by the uneven surface of the moon. You can see bright or dark spots along it when the rising or setting sun catches the peak of a mountain or the high wall of a crater. There is a gradation in size from the maria to large craters, and on to thousands of small craters too small even to be seen in a large telescope.

For decades the maxim has been that the face of the moon is like an open book from which we may read its past history. Since the moon has no atmosphere, there has been no weathering on its surfaces for vast eons of time. For over three centuries this surface has been successfully studied from earth. But these earthly studies are pale in comparison with the spectacular events of the second half of the 20th century. Two of the firsts seemed almost unbelievable to some of the scientists who listened from earth.

Astronomers who have studied the moon for centuries felt frustrated that they could see only one side. The moon turns on its axis in exactly the period that it takes to go around the earth, 27 1/3 days, so it keeps the same side always toward the earth (but not toward the sun). Actually, from earth we can see a little more than 50 percent of the moon's surface because of librations, described as wobblings of the moon on its axis. (On an exam one of my students wrote, "Librations are the warblings of the moon on its axis," to which I replied on the paper, "A librato soprano perhaps?") The librations come from three different sources. The rotation of the moon on its axis is uniform, but its speed in its orbit is not, so features along the eastern and western edges alternately appear and disappear. The equator of the moon is inclined 6½° to its orbit, so the north and south poles are alternately tipped toward and away from us and we can look over the top or bottom. And as the moon is rising or setting we can peek over the edge. From earth we can see 59 percent of the moon's surface at some time, with 41 percent always turned toward us, 41 percent never, and 18 percent alternately toward or away from us.

Most astronomers born in the first half of this century never expected to know what the rear side of the moon is like. But as rockets became more sophisticated and reached farther and farther into space, we began to hope that centuries hence someone would see the far side of the moon.

Then on October 7, 1959, the Soviet probe Lunik III started on its journey behind the moon. On October 24, photographs taken by the space probe on October 18 were released to an excited world. It seemed incredible that earthbound humans were being shown photographs of the side of the moon forever invisible from earth. Since that memorable month other lunar probes, both Soviet and American, have successfully mapped in fine detail virtually the entire surface of the moon, both the near side and the far side.

Following flight after flight by Luniks, Surveyors, Rangers and Apollos, the incredible day dawned: July 20, 1969. For the first time in the 4 billion years of the earth's existence a being from earth set foot on another celestial body. On the Apollo 11 flight, while their companion Michael Collins continued to orbit the moon in "Columbia," Neil Armstrong, followed by Edward E. Aldrin, Jr., stepped onto the lunar surface from the Lunar Excursion Module (LEM) Eagle. "One small step for man — one giant leap for mankind." Photos were taken on the actual surface of the moon with the finest of cameras and film. And from the surface of the moon lunar soil and rocks began to be brought back to earth for intensive and sophisticated analysis.

To those of us who had studied science for half a century, that first step was so fantastic as to seem unreal — a dream from which the earth would awaken. I watched on my TV set beside a western window. My gaze was such that I could see the quarter moon through the window at the same time that I saw the LEM and the astronaut on the TV screen. On one side of my view was the moon as it had been seen since human life began on earth. On the other side was the impossible attained. Which is the greater miracle — that man has reached the moon or that from the earth, a quarter of a million miles away, we could see and hear the landing?

The Surface of the Moon

Most people have now seen TV pictures or close-up photographs of the lunar surface and heard or read the descriptions of the astronauts who have been there. Without an atmosphere, the sky is dark to the edge of the sun, and there is no twilight and no sound. Day and night are each two weeks long, and the temperature range is great, from above that of boiling water (on earth) during the long lunar day to below -130°C. during the long lunar night. There is no water on the lunar surface. And there never has been any.

There are two main types of ground. The rough, broken type is usually bright, sometimes reflecting as much as 18 percent of the sunlight falling on it. This is called the lunar continent, and frequently has a higher elevation than the smooth, dark type, the maria, which reflect about 6-7 percent of sunlight. The continents occupy almost all of the far side of the moon, and the maria are strongly concentrated on the near side, but the most beautiful of all, Mare Orientale, was not appreciated till it was seen in full from the rear. Both continents and maria are covered with pockmarks of various sizes, the craters. The largest craters,

MAP OF THE MOON

South appears at the top.

Figure 9. *Map of moon's surface as seen in telescope. South at top. (From* The Observer's Handbook.*)*

mostly on the far side, are 200-250 miles in diameter. The craters larger than half a mile total more than 300,000 on the near side and several times that on the far side. Most of these craters and craterlets are the result of interplanetary bodies impinging on the moon from the asteroid belt. There are indications that such hits mostly occurred in the first billion years of the moon's existence. There is little evidence on the moon of the kind of volcanic action of mountain building, for example, that we have had on earth. There are, however, domes and wrinkle ridges that are of internal origin. There are clefts and rills, and the bright rays from the largest impact craters, such as Tycho and Copernicus.

The moon's surface is shown to consist of fragmented rock and debris to a depth probably of several miles. This is now considered a normal state for a body like the moon, which does not have an atmosphere to protect it from the inevitable bombardment of rocks and particles from space. This layer is called the regolith. There is evidence for slight changes from time to time in the lunar surface from meteorite impacts, and perhaps from stresses induced when the moon is at perigee (the point in its orbit when the moon is nearest the earth). Perhaps the lunar transient phenomena that have excited the bewildered observers for a century or two are luminous gases that come from vents produced by these stresses.

An important question for the Apollo program to answer was how the chemical composition of the moon rocks would compare with that of the sun, earthly rocks, meteorites and tektites. The answer came out that the moon rocks differ in various ways from all of these others, though the chemical elements are the same, of course. They show no resemblance to tektites, thus ruling out the hypothesis that tektites came from a splash hit on the lunar surface. The moon's rocks contain a higher proportion of titanium than the sun or the earth, but carbon and nitrogen are in short supply; the ratio of iron to nickel is higher than in any other specimen of cosmic matter yet analyzed. The rocks on the moon are all of igneous type, and no sedimentary rocks have been found. An outstanding result of the moon rock analyses is the proof that there never has been any water on the moon.

Once techniques permitted the analysis of moon rocks, the

results were awaited with frantic eagerness. How would they compare in age with those of the earth, or with meteorites? The rocks range in age from about 3,300 to 4,000 million years. The oldest rocks on earth are about 3,600 million years because the earlier strata were consumed in the evolving earth. They are not as old, however, as some of the carbonaceous chondrite meteorites, such as that from the Allende Fall in Mexico which is 4,200 million years old. But the fine dust from the moon is even older, at 4,600 million.

Who's Who on the Moon

The names of the most conspicuous lunar formations on the near side go back about three centuries. As telescopes improved and smaller features became visible, new names were added. In 1936 the British Astronomical Association published *Who's Who on the Moon*, a fascinating booklet listing all the names of lunar features, with translations and derivations. Not all names are as pleasant as Mare Serenetatis, the Sea of Serenity. There is Palus Epidiarum, the Marsh of Epidemics and Palus Putredinus, the Putrid Marsh. Names of earthly mountains prevail in the mountain ranges, such as Alps, Apennines, Caucasus and Carpathians. The names of smaller features like craters are mainly those of earth people, famous philosophers and scientists, with a preponderance of names well known in the 17th to 19th centuries.

With the detailed photographs of the far side, a whole new nomenclature had to be developed for that part. This received final acceptance at the meeting of the International Astronomical Union in England in 1970 when 546 names were added to the 672 already on the moon. Because of the Soviet priority in photographs, some of the main features bear Soviet names, such as the largest sea on the rear side, 190 miles across, Mare Moscoviae. But the new names were selected by a committee that consulted scientific academies of all nations working in astronomy, and they represent deceased persons very prominent in science, philosophy and various disciplines. Six of the new craters were named for Canadian scientists, and 142 for American. The

Canadians for whom craters have been named are five astronomers, C. A. Chant, F. S. Hogg, A. McKellar, R. M. Petrie and J. S. Plaskett and one physicist, J. S. Foster.

Then in 1973 an additional naming was considered necessary by the IAU. The new lunar atlas under construction by the Defense Mapping Agency, Washington, D.C., is on such a huge scale, 1:250,000, that it will consist of 2,300 sheets, occupying more than 700 square yards of paper. Some sheets had no named craters. Accordingly, 50 more names were added, including Sir Frederick Banting from biological science.

The Great Moon Hoax

Just over a century before man landed on the moon, there was about the same amount of excitement over a report that creatures had been discovered on the moon. The story begins in 1835 when Benjamin Day of the New York *Sun* hired for his newspaper a new writer, Richard Adams Locke. Locke, who had come from London, was then 35, and a writer with a real literary gift. Edgar Allan Poe wrote of him, "Everything he writes is a model in its peculiar way, serving just the purposes intended and nothing to spare."

At the time Sir John Herschel, son of the famous Sir William, was in South Africa observing at a new observatory at Feldhausen, near Cape Town. Beginning with the issue of the *Sun* on August 21, 1835, Locke began to tantalize his readers with references to the great discoveries of Sir John Herschel at the Cape of Good Hope "by means of an immense telescope of an entirely new principle." Then on August 25 came three columns to set the world agog. Part of Locke's subtle cleverness was to attribute the article to a Supplement from the Edinburgh *Journal of Science,* which in point of fact had ceased to exist several years before, but readers on this continent would not realize this. With the implication of reprinting from this prestigious-sounding *Journal,* Locke described a great lens 24 feet in diameter, with a magnifying power of 42,000 times, with which Sir John "expressed confidence in his ultimate ability to study even the entomology of the moon, in case she contained insects upon her surface."

The following day the *Sun* carried descriptions of animal life on the moon, with an authentic flavor. "In the shade of the woods on the southeastern side we beheld continuous herds of brown quadrupeds, having all the external characteristics of the bison, but more diminutive than any species of the *bos* genus in our natural history. Its tail was like that of our *bos grunniens;* but in its semicircular horns, the hump on its shoulder, the depth of its dewlap, and the length of its shaggy hair, it closely resembled the species to which I have compared it."

There was an immediate result of this vivid writing on a subject fascinating to all readers. By August 28, 1835, the *Sun,* though not yet two years old, had acquired the largest circulation of any daily in the world. It was selling 19,360 copies compared with a circulation of 17,000 for the London *Times.* The *Sun's* press ran 10 hours a day to keep the public supplied with information on the moon.

The climax of the series, the description of the moon people, came on August 28:

"But whilst gazing upon them in a perspective of about half a mile we were thrilled with astonishment to perceive four successive flocks of large winged creatures, wholly unlike any kind of birds, descend with a slow, even motion from the cliffs on the western side and alight upon the plain. . . .

"We counted three parties of these creatures, of twelve, nine, and fifteen in each, walking erect toward a small wood near the base of the eastern precipices. Certainly they were like human beings, for their wings had now disappeared, and their attitude in walking was both erect and dignified. . . .

"Whilst passing across the canvas, and whenever we afterward saw them, these creatures were evidently engaged in conversation; their gesticulation, more particularly the varied actions of the hands and arms, appeared impassioned and emphatic. . . .

"We scientifically denominated them the *vespertilio-homo,* or man-bat; and they are doubtless innocent and happy creatures, notwithstanding some of their amusements would but ill comport with our terrestrial notions of decorum."

The series concluded on August 31, with a total of 11,000 words. To satisfy readers who asked for back numbers of the series, the *Sun* published a pamphlet, sold at two for a quarter. A set of vivid lithographs illustrating the discoveries was priced at 25 cents.

As these pamphlets were being printed, one Caleb Weeks of Jamaica, Long Island, was leaving for South Africa to acquire some giraffes for his menagerie. When he arrived in Africa, Weeks handed Sir John Herschel one of the pamphlets. It is indeed regrettable that no photographs record the expression on Sir John's face at that moment. That gentleman is reported to have said that he could never hope to live up to the fame heaped upon him, a mild form of understatement.

Back in New York the bubble had burst too. The *Journal of Commerce* in that city decided to reprint the story for the benefit of its readers, and sent a reporter named Finn, who was a friend of Locke, to get the story from him. Apparently friendship triumphed in the ethics involved, for Locke warned Finn not to print the story, saying he had written it himself. The *Journal* did not heed this warning quietly. Instead it blazoned forth the hoax.

As a newspaper coup, however, the success of the story was undoubted. In his book, *The Story of the Sun*, Frank M. O'Brien quoted Edgar Allan Poe: "From the epoch of the hoax, the Sun shone with unmitigated splendor. Its success firmly established the 'penny system' throughout the country, and (through the Sun) consequently we are indebted to the genius of Mr. Locke for one of the most important steps ever yet taken in the pathway of human progress."

So ended the Great Moon Hoax which had an unforgettable impact on North America in the 19th century.

And now we have seen with our own eyes men walking on the moon. But these men came from our own earth. Today astronomers are not hunting for life on the moon. They are quite convinced that no indigenous life exists there, though the first astronauts to return were kept in isolation chambers for a while to be sure that they brought back to earth no harmful forms of life.

But most astronomers are equally convinced that somewhere in the universe there is life and that one of the great pursuits of our time is to find it or establish communication with it. They are not looking for huge winged creatures with impassioned gesticulation. They are looking for the very simplest forms of life — perhaps one-celled animals in the clouds of Jupiter or lichens on the surface of Mars. And they are listening with the huge radio telescopes for a series of signals in a precise artificial arrangement. This would be proof that somewhere far out in the universe intelligent life exists with the technology capable of communicating with us. The proof that there is intelligent life elsewhere in the universe might be the biggest breakthrough in the history of mankind.

★

★

★
FOUR

The Sun

The glorious lamp of Heaven, the sun
The higher he's a getting;
The sooner will his race be run,
And nearer he's to setting.

ROBERT HERRICK, "To
Virgins to Make Much
of Time"

Nothing else in the heavens is so conspicuous as the sun. Its nearest rival, the full moon, gives only about one half-millionth as much light. As a natural consequence of this, the apparent motions of the sun in the sky have regulated the lives of earthly organisms for at least 3,000 million years, and the lives of animals closely related to man for 3 million years.

74

The Sun and the Spinning Earth

Undoubtedly the first astronomical phenomena of which the earliest men were conscious was the rising of the sun in one portion of the heavens and its setting in an opposite part. "It follows as the night the day" has been a built-in part of man's existence. We even tend to assume that elsewhere day and night are similar in duration to our own. And this is rarely true.

Only in the last four centuries, since Copernicus' publication in 1543, have gradually come awareness and agreement that the rising of the sun in the east and its setting in the west are actually caused by the turning of the earth on its axis toward the east.

The diurnal motion (as it is called) makes all the heavenly bodies — not just the sun and moon — appear to rise in the east and set in the west. To be sure, some of the early Greek philosophers like Aristarchus of Samos (310-230 B.C.) preached the idea of a rotating and also revolving earth, but their cries in the wilderness were pretty much unheard for many centuries. It was Copernicus who put the sun in its rightful place, at the center of the solar system, with the planets going around it and the earth rotating on its own axis. But actually he had no decisive proofs for this belief. Modern science has since devised such proofs.

One of the earliest proofs that the actual surface of the earth is turning is provided by the Foucault pendulum, the first of which was set up in 1851 in the dome of the Pantheon in Paris. It can often be seen now in planetariums or science museums. As the pendulum swings, the ground skews under it because the southerly end of the sand table is rotating at a more rapid rate than the northern (in the northern hemisphere).

Then there is the Coriolis effect. This is a sidewise drifting of objects moving over the surface of the earth, to the right in the northern hemisphere. The difference in speed with difference in latitude causes a moving object to skew around, just as the pendulum does in a smaller locale. The Coriolis effect increases markedly with the velocity of the moving object.

One of the earliest times it was taken into account in an important way occurred in World War I when the German Big

Bertha shelled Paris from a distance of 70 miles. Experts manning the gun allowed for a Coriolis drift of a full mile. With the speed of jet planes, the effect is even stronger.

In ordinary living we are moving too slowly to notice it, but its effects are present. Actually it is strong enough to shift an auto traveling at 60 miles an hour 15 feet a mile to the right, but the friction of the tires holds the vehicle to the road. It gives bath drains a counterclockwise whirl in the northern hemisphere, and clockwise in the southern. It may explain the tendency of lost explorers to walk in circles, toward the right at the north pole and toward the left near the south pole. Perhaps the Coriolis effect causes the penguins in the Antarctic to waddle toward the left! And another suggestion is that the direction of flow of the fluid in the semicircular canals of the ears of birds may be a factor responsible for their amazing migratory ability.

Time Determination

Before the dawn of recorded history, mankind recognized that the position of the sun in the sky is an indication of the progress of the day. And as man's knowledge of the earth and its relation to the sun increased, so did the sophistication of the devices man uses for timekeeping. Alfred the Great used "good tall candles," divided into equal segments. As they burned, they showed the time of day or night. The first clock in the modern sense was constructed by Huyghens, who used Galileo's discovery of the pendulum for it. Nowadays it is a commentary on the complexity of our civilization that relatively few persons (who have not had a course in astronomy or navigation) can tell you in specific terms the scientific facts behind the time that their watches are carrying. And yet these facts are actually very simple.

For centuries the position of the real sun in the sky was used for the time. This is called apparent solar time. It is the hour angle of the sun in the sky, that is, its distance in time from the meridian, the great circle through the zenith and the north and south points of the horizon. When the sun is on the meridian it is "high noon," because that is the highest altitude the sun will achieve on that particular day. Its altitude increases until it reaches the meridian. Once it crosses the meridian the sun steadily loses altitude.

It is a very easy thing to set up a stylus or similar device called a gnomon on the north-south line to show by its shortest shadow when the sun is on the meridian. The next step is to make a graduated dial around it, reading hours and minutes east or west of the meridian. And then you have constructed a sundial that reads the apparent solar time at the place you have set it up, if you have done it properly. On February 11, 1974, artist Yuri Schwebler turned the Washington monument into the largest gnomon on record. When snow was on the ground, he instructed a snowplow to make straight paths in the proper positions to indicate hours as the shadow fell on them.

There are many types of sundials in existence — horizontal, vertical and equatorial. The simplest kind to make is the horizontal, on which the hours are marked on a flat, horizontal plate. If you set one up yourself, you must be aware of certain requirements. The stylus of the dial, the thin sheet that casts the shadow, must be tilted at an angle equal to the latitude where it is to be used. (This is where many people go astray, particularly antique hunters, who enjoy bringing dials back from the British Isles, from latitude around 54°, and setting them up perhaps in Washington, latitude 39°). Then the stylus must be exactly oriented in the north-south line and the dial plate level. It is not

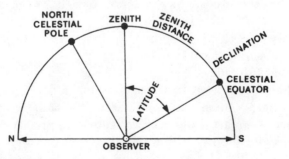

Figure 10. *The altitude of the celestial pole above the horizon is equal to the latitude of the observer.*

difficult to make a dial that will be accurate to about five minutes of apparent solar time.

Sundial time was fine in the early days of civilization, but as watches and clocks were developed it became obvious that they did not keep apparent solar time. They were moving in a uniform manner and apparent solar time was not. Actually the motion of the real sun is not uniform enough to make it a precise timekeeper each day, for two reasons. It does not move uniformly in the plane of the earth's equator, and its daily motion varies with the earth's varying distance from aphelion to perihelion. Watches and clocks are really reading the kind of time that the sun would keep if its motion were uniform, that is, mean solar time.

For every day in the year there is a numerical relation between apparent and mean solar times, and various almanacs publish it annually. The difference, apparent minus mean solar time, is called the equation of time. The maximum difference between the two, 16 minutes, comes in the first week in November when the real sun is that much ahead of the mean. The maximum difference in the other direction comes the second week in February, when the real sun is 14 minutes slow. Four times a year, the difference between the two is zero, around the middle of April, the middle of June, the first week in September and the last week in December. The authoritative *Nautical Almanac and Astronomical Ephemeris of England,* beginning in 1767, used apparent solar time and did not change from it until 1834, on the advice of the Royal Astronomical Society. Even then the President of the Royal Astronomical Society, Sir James South, was against the change to mean solar time because he considered it alien to astronomical thinking.

For many decades mean solar time was used successfully, but with the advent of rapid transportation, difficulties increased. When transportation was by horse or by sailboat it was not necessary — in fact it was very difficult — to predict the arrival time on a long journey within a few minutes. But the Iron Horse changed all this. Once it began to make transcontinental journeys with many stops along the way, timetables expressed in mean solar time became a gigantic headache. Mean solar time depends not only on the mean sun but also on the longitude of the observer. A difference in longitude of 1° means a difference of four minutes in time. So the timetables for a rail journey involving hundreds or thousands of miles, with many stops, became a nightmare of confusion.

This was what led to the idea of Standard Time. Sir Sandford Fleming in Canada and Dr. Charles Dowd in the United States were largely responsible for this idea, which was adopted at a longitude conference in Washington in 1884. Standard Time has been a great boon ever since. The world was divided into time zones an hour apart. In these Standard Time zones, the meridian adopted every 15° runs a straight course, but the boundaries of the zones are often irregular, to conform with local geography and commerce. For example, two cities close together, doing business with one another, will probably be in the same time zone even though by strict arithmetic the time-zone boundary would fall between them. The time-zone boundaries over North America are shown in the map on page 80. .

You can see why, when you compare your watch which is reading Standard Time, with a sundial reading apparent solar, they may not agree! An uninformed observer is apt to blame the sundial as an old-fashioned instrument, and thinks it fortunate that the dial is superseded by modern watches. To make a valid comparison you must take account of two corrections, one for the difference between apparent and mean solar time, and the other for the longitude of the dial. An example which works out this problem is given in Appendix III. Actually, even in modern times a sundial is a very fine type of instrument to tell time within five minutes or so. Its true fault is one that is occasionally found as a motto engraved on the dial, *"Horas non numero nisi serenas."* ("I tell no hours but sunny ones.")

The 20th-century improvement on Standard Time has been the introduction of Daylight Time, which in many sections of North America is taken on in the spring and given up in the fall. It is one hour ahead of Standard Time. To remember which way to change your watch, use the adage, "Spring forward, fall backward." There is always discussion, even argument, as to the value of Daylight Time, and the reason is obvious. Its advantage

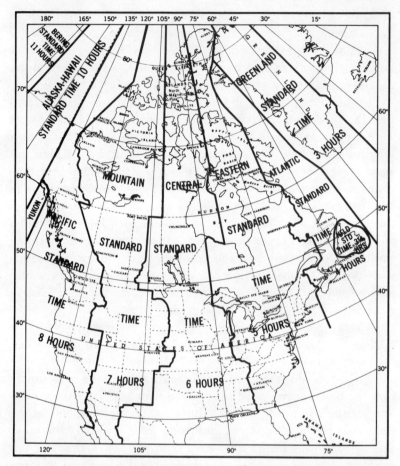

Figure 11. *Map of Standard Time Zones. (Produced by the Surveys and Mapping Branch, Department of Energy, Mines and Resources, Ottawa, Canada 1973.)*

varies with the position of the locality with respect to the adopted meridian. A community that is 30 minutes west of its Standard Time meridian is using time that is always half an hour ahead of its own local mean time, and is understandably reluctant to increase the difference to one and a half hours. Conversely, communities far east of their time meridian are using Standard Time, which is behind their own local time, and don't object to having it quickened up a bit. During World War II in Britain an extra hour was added onto Daylight Time as a war measure, and

Hitler was blamed for this inconvenience. One man who was asked for the time replied, "One by God's time, two by man's time, and three by the Devil's time."

Since the era of radio and TV began, it has become virtually no problem for a household to have accurate time, and we now take it more or less for granted. Time is determined at certain observatories of the world and then widely distributed by broadcasts. Universal Time is the Standard Time of the meridian of Greenwich, on a 24-hour basis. It is five hours ahead of Eastern Standard Time. Ephemeris Time (which has a cumbersome definition) is the same within half a minute. (The difference between Ephemeris Time and Universal Time is given each year in *The American Ephemeris and Nautical Almanac*.) It runs on uniformly and is an invariable unit of time.

In 1965 the Bulova Watch Company conducted an experiment in North Conway, New Hampshire (pop. 1,100), to show how time-dependent our civilization has become. In "Project Time Out" the faces of all municipal clocks as well as those in offices and stores were covered with cards, which showed only a question mark. The inhabitants found it "nerve-wracking" to live by guesswork. For example, sulfurous fumes resulted when one housewife inaccurately guessed the time for eggs she was cooking on the stove.

However, a good many grade-school children "beat the rap" by being taught the "woodsman's timepiece" by their teacher. A person stands back to the sun, with one arm straight forward and the thumb pointing up vertically. If you have some idea where the north-south line is (it can be determined from the shortest shadow) then the angle that the shadow of your thumb makes with this line will give you some indication of the time of day.

A place where time can not merely stand still but can even turn backward is the International Date Line in the Pacific. Because time gets progressively earlier and earlier as you go round the world toward the west, there has to be a day of reckoning. That day is approximately at the 180th meridian, a place so chosen as to be mostly over water, and hence less apt to cause confusion in the lives of many persons. As you cross the line from North America to the Orient, the date suddenly becomes one day later. As you cross from the Orient to North America the day is repeated; that is, if you cross at midnight Sunday you have Sunday all over again. This situation led to a delightful Victorian era conundrum. What is the maximum number of Sundays in February? The answer is 10. The conditions, however, are somewhat restrictive. The maximum number can be achieved only in a leap year when February 1 falls on a Sunday. And only on a vessel which makes weekly sailings from Siberia to Alaska, starting on February 1.

The Sun and the Revolving Earth

When they followed the daily course of the sun across the heavens thousands of years ago, early peoples also noticed that the points of rising and setting change appreciably from week to week, shifting toward their southernmost position and then back toward their northernmost in about 365 days. Also, the highest altitude of the sun when it is on the meridian changes. It is one of the many ironies of our civilization that primitive men were more conscious of this shift than are our highly educated, 20th-century humans, hemmed in by city skyscrapers. By blocking the view, the buildings and trees of a large city often render the change in

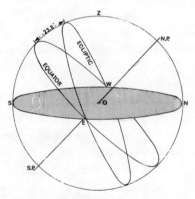

Figure 12. *The annual path of the sun, the ecliptic, is a great circle inclined 23½° to the celestial equator. 0 is the observer at center of celestial sphere.*

position of sunrise and sunset points so difficult to observe that many persons go through life relatively unaware of them.

If you want to follow the sun's position, even if you do not have a good east or west horizon, but have a south wall in which you can make a tiny opening, you can use an ingenious method described by Dr. Richard M. Sutton in *Physics Today*. He made a hole one centimeter in diameter in an opaque sheet of aluminum at a height of 140 centimeters above the floor. The sunlight passing through the hole makes a spot whose position may be marked precisely to within one millimeter. The north and south swing of the sun and its deviation at noon from the meridian line make a pattern that is fascinating to follow.

Just as the sun has an apparent daily motion across the heavens because of the earth's rotation, so it also has an apparent annual path around the sky because of the earth's annual revolution around it. By being so bright, the sun itself makes it difficult to observe its path among the stars. When it is above the horizon, virtually every star in the sky has faded out of sight. Therefore it is not easy to detect that the sun is moving slowly eastward among the stars, steadily, without interruption, by 1° a day, an amount equal to twice that of its apparent diameter.

This motion is further masked by the fact that the rotation of the earth on its axis swings the whole sky westward at a far faster rate than the sun crawls east. Therefore we are conscious of the wide daily sweep of the sun across the sky to the west, and little conscious of its small daily eastward creep. If in its blaze of glory the sun did not blot out all the stars around it, you would be fascinated to watch the easterly course it makes.

You can watch it, however, in a backhanded fashion, by observing the westward march of the constellations. This daily motion of the sun 1° toward the east makes individual stars rise four minutes earlier each night, and also set four minutes earlier. Thus constellations that rise in the east at 7 P.M. on September 1 will rise about 5 P.M. on October 1. Correspondingly, they will set two hours earlier in the west. During the year, then, we have a steady westward march of the constellations. Those that are visible in the warm summer evenings — the beautiful star clouds of the Milky Way along the southern horizon — are gone when winter comes. They are replaced by the winter constellations like Orion, with its two 1st-magnitude stars, red Betelgeuse and blue Rigel.

The apparent annual path of the sun, the ecliptic, is a great circle inclined $23\frac{1}{2}°$ to the equator. It is this inclination that gives such a wide swing to the sunrise and sunset points as well as to the meridian altitude of the sun. The sun's motion is always easterly, but combined with either a northward or southward motion according to the seasons. The points where the sun is farthest from the celestial equator — $23\frac{1}{2}°$ from it — are the solstices. The word "solstice" means literally "the sun stands still." Near the solstice the sun has little but eastward motion and hence appears to stand still in its rising or setting point. The two equinoxes occur when the sun is on the equator. Again the name is derived from the fact, because day and night are then equal on that day the world over. The four points — the vernal equinox, summer solstice, autumnal equinox and winter solstice — are points in both time and space. Each one has a date (which depends on the individual year) and a position among the stars, which varies slightly each year and markedly over a long period of time.

Centered around the ecliptic is the zodiac, a band of the sky 16° in width which always contains the sun and moon, and most of the time the naked-eye planets. Venus, the nearest to us, can sometimes slip off the edge. With 12 constellations in the zodiac, the sun spends about a month in each. The famous angler Isaak Walton constructed a rhyme of the names of the constellations of the zodiac which makes it easy to remember them. One version goes like this:

> *The Ram and Bull lead off the line*
> *Next Twins and Crab and Lion shine*
> *The Virgin and the Scales:*
> *Scorpion and Archer next are due*
> *The Goat and Water Bearer too*
> *And Fish with glittering tails.*

Neolithic Monuments

The important prehistoric monument Stonehenge was obviously based on the apparent annual path of the sun. The massive stones, rising above the Salisbury Plain, cast a spell of awe and mystery on the beholder. Dr. John Smith in 1771 first drew attention to the astronomical orientation of the stones. If a spectator stands in the center of the horseshoe, he will see the sun rise over the Friar's heel at the summer solstice. So keen have been the interest and studies in this relic of former times that more than a thousand papers or items have been written about it. In the last decade, however, modern technology has entered the scene and giant computers have been successfully used to study the problem of the stone positions.

It was Sir Norman Lockyer, the noted astronomer, in 1901 who first made some crucial observations from this remarkable alignment of massive stones and from them drew deductions as to the probable dates when this monument was constructed. Lockyer considered the axis of Stonehenge to be aligned accurately by its builders upon the point of midsummer sunrise at the date of its construction. The position of this point depends on the angle the ecliptic makes with the celestial equator, called the obliquity of the ecliptic. Since this changes over the centuries, Lockyer could use this to determine the period when Stonehenge was built. His computation for the date of construction of the sarsen (sandstone) circle was 1680 B.C. $\pm$ 200 years. Since his time the obliquity of the ecliptic has become known with higher precision and his date has been revised to 1840 B.C. $\pm$ 200 years. Later, R.S. Newall suggested that the alignment was made on the midwinter sunset to the southwest, rather than on the midsummer sunrise to the northeast.

Though modern scientific methods of higher precision now quibble a bit about Lockyer's findings, the fact remains, however, that his determination of the date is in very close agreement with those dates of other independent methods. These come from a study of the secondary neolithic cultures, the Beaker and Wexhill, which were behind the construction of this great edifice. And radiocarbon dating has confirmed the great age of

Stonehenge. Some charred oak was unearthed in previously untouched areas under Stonehenge in 1950 by Stuart Piggott of the University of Edinburgh. Professor Willard Libby of the University of Chicago measured the age of this precious old dust in his laboratory as 3,800 years with an uncertainty either way of 275 years.

And now some of the most powerful instruments of the 20th century, the great electronic computers, have been used to study the alignment of these stones. With an IBM 7090 and 7094, Dr. Gerald Hawkins of the Smithsonian Astrophysical Observatory tested stone alignments. He computed horizon positions for the rising and setting of the sun, moon, planets and some bright stars, using a construction date of 1500 B.C. He detected no correlation for stars or planets, but a remarkable correlation appeared with the positions of the sun and moon in midwinter and midsummer. The probability that these positions come from chance alignment is less than one in a million. Hawkins calls Stonehenge "A Neolithic Computer," and postulates that it was used for prediction of eclipses.

Less than two minutes of time on the giant modern computers was needed to show the reasoning behind the giant neolithic computer. The contrast between the centuries of construction, the millennia of mystery, and the seconds of realization of this aspect of the astronomical significance of this mighty monument is indeed dramatic.

Stonehenge is only one of many neolithic monuments in the British Isles and Europe, but it is the best known, and probably the most elaborate. Dr. A. Thom of England estimates that on the west coast of Scotland alone there are 300 of these early stone markers. Many of them indicate the position of the setting sun at the equinoxes or the solstices. Through the obliquity of the ecliptic, many of these constructions may be dated as around 1800 B.C.

North America may have its counterpart, though at a date three millennia later, in a Woodhenge. A prehistoric site near St. Louis, Missouri, supported a flourishing Indian civilization about 1000 A.D. Recent excavations at "Cahokia" have uncovered four huge circles of wooden posts which were carefully spaced. These are near "Monk's Mount," the largest prehistoric earthwork in North America, investigated by the Cranbrooke Institute of Science.

This Woodhenge was 410 feet in diameter, and a very precise circle. According to Dr. Warren Wittry of the Institute, it was laid out with the use of a peg and rope compass. One post seems to be in a key position. An observer at it in 1000 A.D. would see the sunrise at the summer solstice in line with one of the henge posts.

In the past decade, researchers have found that large stones on Mystery Hill, North Salem, New Hampshire, seem to have astronomical alignments and date back to 2000 B.C.

The Year and the Calendar

It is the return of the sun to the vernal equinox that gives us the important interval called the tropical year. After one tropical year has passed, the seasons repeat. In its use in a calendar this interval has regulated the lives of much, but by no means all, of mankind for at least 2,000 years. Two principal calendars have been in use for many centuries. One is based on the sun's annual path; the other, on the moon's monthly path — in particular, its phases. Some early calendars tried to take both into account. But obviously a good calendar must contain an integral number of days, and the three quantities — the length of the day, the length of the year and length of the synodic period of the moon (the period of its phases) — are incommensurable. No calendar can satisfactorily combine them all. It is impossible to construct a calendar based on the year that will always have the new moon on the first day of the month. Accordingly, in our western calendar we have dropped the moon except in one regard: setting the date of Easter. Here the old influence of the phases of the moon is still left in the calendar, for Easter is the first Sunday after the 14th day of the moon (the full moon) after the vernal equinox. The result of this combination of the lunar and solar influence is a most inconvenient swing of five weeks in the date of Easter. This swing has no religious significance, it does no one any good and it causes inconvenience to millions. "Moon-wandering Easters" astronomers call them. The dates for Easter for the rest of this century are given in the following table.

Dates of Easter as derived from Augustus De Morgan for Rest of 20th Century

Year	March	April	Year	March	April
1977		10	1989	26	
78	26		90		15
79		15			
80		6	91	31	
			92		19
81		19	93		11
82		11	94		3
83		3	95		16
84		22	96		7
85		7	97	30	
86	30		98		12
87		19	99		4
88		3	2000		23

Our calendar is based on the numerical value of the tropical year, 365 days, 5 hours, 48 minutes, 46 seconds. Julius Caesar with the help of the astronomer Sosigenes did a splendid job of calendar reform in 45 B.C. when the value of the tropical year was even then fairly accurately known. Caesar and Sosigenes had the acumen to disregard the phases of the moon, and they solved the problem of the extra hours over integral days in an ingenious way. Now 365 days and 6 hours (to which you might round off the tropical year) is 365¼ days. So Caesar and Sosigenes invented the leap-year principle. Normally the year would be 365 days long, but in every fourth year an extra day would be added. In that way the *average* length of the year would be 365¼ days, and the value of the tropical year would be satisfied with an integral number of days each year.

However, 365¼ days is just a little too much — 11 minutes more than the precise value of the tropical year. During the lifetime of one emperor this amount would not matter, but in the course of centuries it adds up. By medieval times the vernal equinox had moved forward to March 11 from its earlier date of March 21. The Julian calendar was adding three days too many in 400 years. Eventually of course, the seasons would get all out of kilter. Summer would come in winter and vice versa. So in 1582 Pope Gregory XIII instituted another calendar reform, with the help of the astronomer Clavius, who had devoted his working life to considerations of the calendar. They introduced a very clever revision to the leap-year principle: they decreed that century years not divisible by 400 would not be leap years. This was a painless way of getting rid of the three extra days in 400 years. The Gregorian calendar is the one we use today.

Their other move, however, was not so painless. They dropped 10 days from the calendar in order to set the vernal equinox back to March 21, where it had been earlier. October 4, 1582, was followed by October 15. This calendar change met with stiff opposition from the non-Catholic world. The Catholic countries followed the Pope's dictum promptly. The Protestant countries took centuries in some cases to adopt the new calendar.

Not until 1752, when the differences between the two calendars had grown to 11 days, did the British Dominions adopt it. Lord Chesterfield introduced the bill to the British Parliament, with the backing of Martin Folkes, President of the Royal Society, and James Bradley, Astronomer Royal. The act was carefully framed to prevent injustices, such as landlords collecting rent for the missing days. Nevertheless, there were riots in the streets of Bristol, with people shouting "Give us back our fortnight." Something no act of Parliament could counteract was the firm conviction many persons held that, as a result of this change, their lives would be shortened by 11 days. Would you like to try to explain to an uneducated man in the street that the next day would not be September 3, but would be September 14?

Another item beyond the control of Parliament was the famous Glastonbury Thorn which for years had bloomed on Christmas

Day. It is not recorded what helped the Thorn to make up its mind, but to the relief of all, the Vicar of Glastonbury was able to announce to an anxious world that the Thorn in 1752 blossomed on Christmas Day New Style!

Precession of the Equinoxes

The time that the sun takes in its apparent course to go around the ecliptic and back to the vernal equinox is not the same as it takes to return to the same star in the heavens. The latter interval is the sidereal year, 365 days, 6 hours, 9 minutes, making it approximately 20 minutes longer than the tropical year. The reason for the difference is that the equinoxes are moving among the stars, slipping westward by 50 seconds of arc a year. The sun gets back to the vernal equinox 20 minutes sooner than it returns to a fixed star. The 20 minutes is the time it takes the sun to traverse the 50 seconds of arc.

This effect is known as the precession of the equinoxes. The equinox has literally preceded, that is, stepped forward to meet the sun. This was discovered by Hipparchus about 150 B.C. when he noticed that over an interval of many years the dates when a star rises or sets with the sun were slowly changing. The effect is easily understood, but the term itself sometimes bewilders people, as Rudyard Kipling when he wrote in *The Jungle Book*, in the story of the Elephant's Child: "In the middle of the precession of the equinoxes when the equinoxes had preceded according to precedent . . ."

This precession of the equinoxes is caused by the pull of the sun and moon on the equatorial bulge of the rapidly rotating earth. You can easily see the effect for yourself if you spin a top and then by a slight pressure of your finger deflect it from the vertical position. You can then watch its axis describe a circle as the top spins. In the case of the earth, whenever the circle is completed, another 26,000 years have passed.

One result of the precession of the equinoxes keeps astronomers busy. The coordinates for the positions of heavenly bodies are computed from the position of the vernal equinox, which is

constantly changing. The celestial coordinate corresponding to longitude is called right ascension and is the angular distance of a body measured along the celestial equator eastward from the vernal equinox. (The celestial equator is the projection of the earth's equator on the sky.) Declination is the term for the celestial coordinate corresponding to latitude and is the number of degrees north or south of the celestial equator. Star catalogues, often involving hundreds of thousands of stars, have to be recomputed when the change becomes significant. A large telescope cannot be set properly on a star unless the equinox used is not more than several years from the date. Catalogues always state the year of position of the equinox they choose.

Perhaps the most interesting effect of the precession is the wandering of the celestial poles, which is discussed in Chapter 8.

The Sun, Our Nearest Star

The sun is an intensely hot globe of gas under high pressure, at a distance of approximately 93 million miles from us, with a diameter of 865,000 miles. The temperature of its surface is 6000°C., but that of its interior is about 14 million°C. Atomic energy, with hydrogen atoms combining to build up helium, and mass converted to radiation, keeps the sun shining. But its luminous outpouring destroys 4 million tons of its mass every second.

The sun is the one celestial body you must not look at fixedly under normal conditions. The surface brightness of the sun is so great that even a small portion of it, as at a partial eclipse, can produce eye damage. Oculists even in the 20th century say that they see many cases of serious eye damage resulting from a fixed look at the sun. You can look at the sun through a telescope only if it has been fitted with a solar eyepiece or special filter.

Most of the time, to the unaided eye, the sun appears as a featureless globe, too bright to study for detail. Under certain natural conditions you can see structure, when haze or thin cloud cuts down the glare and you can look safely at the surface. This most frequently occurs when the sun is low in the sky, near its rising or setting points, and its rays are coming through the greatest depth of atmosphere.

Perhaps early man noticed the dark spots on the sun as much as 4,000 years ago, because celestial observations were made and recorded in China very early indeed. These sunspots are huge, magnetic storms on the surface of the sun, appearing black by contrast. Despite their high temperatures, of 4800°C., they have been called the greatest refrigerators known, because they maintain a temperature 1200°C. lower than their surroundings. Sometimes spots with an area of millions of square miles are large enough to be seen by the naked eye here on earth, 93 million miles away. Long before the invention of the telescope, the Chinese made many records of spots seen with the naked eye. For example, in the Encyclopedia of Ma Twan Lin on April 4, 355 A.D., a spot is recorded "like a peach"; on September 8, 359, "like a hen's egg"; on March 29, 370, "like a large plum"; and on April 6, 374, "like a duck."

However, it is not necessary to wait until nature provides an eye screen in the form of haze or cloud for studying the sun's surface. You can embark on a do-it-yourself project with a pinhole camera. Experiment by punching a tiny hole in cardboard and projecting the image onto a plain, clear surface. Frequently around us there are materials that automatically act as such a camera, if you are alert to them. For example, at the David Dunlap Observatory there are long Venetian blinds over large, high windows in the entrance hall. When in the afternoon the sun is in the western sky, its light falling through the holes in the blind projects images on the floor some 25 feet below. If there is a sunspot large enough to be seen without a telescope, it shows up well on the images of the sun on the floor.

Although sunspots have been known for many centuries, it is only since the invention of the telescope that observation of them has been systematic. Galileo in 1610 with his first telescope made many drawings of them. Curiously, one of the most important facts about sunspots was discovered only a century and a third ago. This is the sunspot cycle: the numbers of spots increase to a maximum and then drop down to a minimum in a period of approximately 11 years. It was discovered in 1843 by that indefatigable observer Heinrich Schwabe of Dessau. In awarding

Schwabe the gold medal of the Royal Astronomical Society in 1857 for his discovery, the president of the Society commented, "For thirty years never has the sun exhibited his disk above the horizon of Dessau without being confronted by Schwabe's imperturbable telescope, and that appears to have happened on an average about 300 days a year!" Anyone with experience in astronomical observation can appreciate what this means — that seven days a week for 30 years Schwabe would be close to his telescope. If cloudy, he would stay with it in case the sky cleared enough for him to get an observation.

Ever since Galileo's records, there had been fairly systematic observations of the sun. Why then was this cycle missed until the middle of the 19th century? Mainly because in the last decades of the 17th century the sun's activity had dropped to a strangely low level. For months on end not a spot was visible on the sun. Although telescopes were then rather crude, they cannot be blamed for the sparsity of spot observations. The spots simply did not then exist.

Some decades after Galileo first observed sunspots, the sun went into a long quiet stage that has never yet been repeated. As the 17th century wore on, spots became scarcer and scarcer. For about 70 years a remarkable period of quiescence prevailed on the sun. A single spot occurring in this interval was worthy of published notices. And yet from the years when an occasional spot did appear, one can trace back a weak 11-year spot cycle.

But the sunspot cycle has far more influence on our earth than just providing sun watchers with a fluctuating number of spots. The correlation between the sunspot cycle and magnetic disturbances of the earth's field is very strong. And we have already seen the strong correlation with the aurora. In its quiet stage the sun does not seem to send out the streams of high-energy particles that set off a brilliant discharge in the earth's upper atmosphere. There are indications that periods of cold climate on earth correlate with low solar activity. Since the earth has had a relatively active sun for the last two and a half centuries, we may well ask ourselves if we know what our climate would be like should the sun lapse again into many decades of quiescence.

Actually, modern techniques show that the magnetic polarity

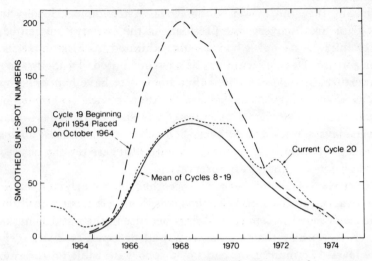

Figure 13. *The recent sunspot cycle, with Cycle 20 approaching minimum. (From* The Observer's Handbook.*)*

of the sunspots reverses at the end of each 11-year period, so it takes about 22 years for conditions on the sun to repeat. Furthermore, there is evidence of even longer periods in solar activity, running to many decades. These lead to abnormally high spot maxima or low minima in the 11-year cycle. The maximum of the current cycle, number 20, was reached in the winter of 1968-69, with a small secondary maximum in early 1972. Probably 1975 will prove to be the minimum year. The spots of the next cycle have already begun to appear with reversed polarity in higher latitudes.

The solar cycle also shows up in the appearance of the outer parts of the sun, the brilliant red prominences and the silvery white corona. These can be seen with the naked eye only at a time of total eclipse of the sun, and are discussed in the next chapter on eclipses. Sophisticated instruments have been developed, including a coronagraph, which permit the study of these features from observatories at high altitudes. The corona itself seems to have a very long extension out from the sun which can be seen under ordinary conditions, and this is known as the zodiacal light.

"Before the phantom of false morning died," wrote the Persian astronomer-poet Omar Khayyam eight centuries ago. Annotated editions of his famous *Rubaiyat* explain that about an hour before dawn in the tropics the eastern sky brightens in a kind of false dawn (Subhi Kházib).

This is the effect of zodiacal light. The name comes from the position of this glow. The zodiacal light is a cone-shaped band of light, with faintly iridescent colors, centered on the ecliptic, and hence in the zodiac. It is comparable in brightness with the Milky Way. Persons living in the tropics, especially in arid regions or at high altitudes, can hardly miss its brilliance. In the temperate latitudes, where most of us live, the zodiacal light is not usually prominent enough to attract attention.

But at certain seasons of the year it can be seen easily if you look in the right place at the right time. In the north temperate zone it shows up best in the spring in the western sky after sunset, or in the autumn in the eastern sky before dawn. The reason for this lies in the high angle which the ecliptic makes with the horizon at these times.

Look for the zodiacal light in February, March or April on a clear moonless night, an hour or so after sunset. As soon as the sky is dark you will see the cone-shaped band of light extending up from the horizon. Probably you have already seen it sometime and thought it was just sunset afterglow. Its height depends on the interval after sunset. As much as two hours after sunset it will usually be too near the horizon to be visible. Do not confuse it with the Milky Way, which, during those same months in the early evening, comes down to meet the horizon in the northwest. You should be able to see a dark band of sky between the zodiacal light and the Milky Way.

And what is this misty band? It is sunlight reflected from myriads of particles concentrated toward the plane of the earth's orbit and scattered between the earth and sun. It is a continuation or extension of the sun's corona that you saw at total eclipse if you were lucky. Beyond a million miles from the sun's center the coronal light comes chiefly from the scattering of sunlight by small particles. At the 1973 eclipse the corona was traced out more than 15 solar radii by Dr. William Liller of Harvard at Lake Rudolf in Kenya. This outer faint corona is considered the innermost part of the cloud of particles responsible for the zodiacal light. By estimate particles 1/25 inch in diameter and five miles apart will account for the brightness of the zodiacal light. In the vast volume of space over which these particles are spread, this adds up to a lot of material. Doubtless much of this dust has been shed by comets as they wheel in and out from the sun.

The zodiacal light is important. Nowadays highly technical observations of it are being made. From aircraft and satellites and space expeditions, from the earth's surface at high altitudes, as at 17,630 feet from Mount Chacaltaya, Bolivia, the information obtained will aid in our understanding of the false dawn.

Stars, including our sun, form out of a cosmic cloud of dust and gas, an enormous cloud some light years in diameter. This breaks up into fragments perhaps three light years across. Each fragment then condenses down to a body some 10,000 astronomical units in size. This is called a "protostar". Dark globules seen against star clouds may be such protostars. The protostar then heats up by contraction and begins to shine. Only a short span of time, up to 100 years, is necessary for the star to become visible after the gravitational collapse of the interstellar cloud. A star's lifetime may extend to tens of billions of years.

The course the evolution takes depends to a great extent on the mass of the star, a fundamental characteristic that varies greatly from star to star. The mass will determine how much nuclear fuel is available to it. There is a wide range in the physical properties of stars, but our sun may be considered an average star. It is now some 5 billion years old. In another 5 billion years it will have consumed so much of its hydrogen that it will evolve to the red giant type of star. Its diameter will increase to 250 times the present value. The expanding sun will absorb the planets Mercury and Venus and probably the earth. In this expansion the sun's density will become only a tenth that of air, compared with its present average density, a fifth that of the earth.

As the nuclear fuel becomes mostly consumed — helium and higher elements after hydrogen — the sun will contract to 1/100 of its present diameter, similar in size to the earth, and its density will become enormous. It will be a white dwarf. Its substance may have a density 200,000 times that of water, like Sirius B, the famous white dwarf companion to Sirius, the brightest star in the sky. Density like this is unattained by any object on earth.

Stars starting originally with masses much smaller than our sun shine only on the light generated by gravitational contractions. They do not have enough mass to start their nuclear furnaces going. Those with less than a tenth the sun's mass evolve to black dwarfs. Their light is so feeble that only a very few have so far been observed, though there must be many around.

Stars much more massive than the sun may experience pulsations along their evolutionary path, as discussed in Chapter 9. They may evolve to a different endpoint, that of a neutron star or even a black hole (Chapter 10).

★

★

★

FIVE

Eclipses

Since before the dawn of history, man has regarded a total eclipse of the sun as one of the most awe-inspiring phenomena of the skies. Anyone who has seen the full beauty of a total solar eclipse can readily understand this feeling. And the knowledge that modern astronomy has brought us in no way diminishes the magnificence of the spectacle.

Eclipses of the Sun

When conditions are just right at a new moon, the dark disk of the moon passes centrally across the bright disk of the sun and for an interval never exceeding 7½ minutes may completely blot it out. We earth dwellers are fortunate that the apparent diameter of the moon so closely balances that of the sun that it makes total solar eclipses possible. Were the moon's diameter only 140 miles smaller, we could never see a total eclipse of the sun from the earth!

Total eclipses of the sun are very rare. If you stay in one spot on the surface of the earth and wait for one to come to you, the chances are that one won't. On the average, a total eclipse track crosses a given spot on earth once in 360 years. Naturally averages, like New Year's resolutions, can be broken, and the same place can have two total eclipses in a very few years. The tracks of the total eclipse of March 7, 1970, and that of July 10 1972, were common to one small part of Nova Scotia. It was the Chinese who kept the earliest and most diligent records of eclipses. Some are in the *Su Ching,* the *Book of Historical Documents.* There is some uncertainty about the earliest recorded eclipse. According to Oppolzer, it is probably that of October 22, 2137 B.C. He made a memorable computation of all solar and lunar eclipses visible the world over from 1208 B.C. to 2163 A.D. The requirements for astronomers were fairly stringent in the days of the early Chinese chronicles, and would discourage all but intensely devoted members of the profession. In the prediction of eclipses, "Being before the time, the astronomers

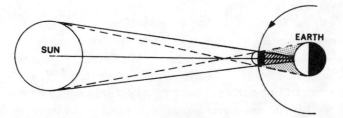

Figure 14. *A solar eclipse occurs at new moon when the moon's shadow touches the earth.*

are to be killed without respite; and being behind the time, they are to be slain without reprieve." Two royal astronomers, Hi and Ho, were apparently suitably punished for their failures with respect to an early eclipse. The first certain eclipse date from Chinese annals is September 7, 776 B.C., more than 1,000 years later than the earliest possible record. This eclipse followed one of the moon on August 21, making a double date that is unequivocal.

Another ancient civilization, that of the Egyptians, is totally lacking in recorded dates of early eclipses. The reason for this is puzzling, and the fact itself is a hindrance to the dating of Egyptian dynasties. And yet there is evidence that the Egyptians witnessed total solar eclipses, because the symbol they used for the sun god is a circle with striated wings to right and left, resembling the form of the solar corona at sunspot minimum.

Babylonia and Assyria, though, were rich in early eclipse records. The Chaldean astronomers zealously observed eclipses. The earliest recorded, with Nineveh just outside the path of totality, was a morning eclipse on June 15, 763 B.C. Experts think this eclipse, which would be visible also in Samaria, is that mentioned in the Bible, in Amos 8:9: "I will cause the sun to go down at noon, and I will darken the Earth in the clear day."

Another famous eclipse is that which occurred on May 28, 585 B.C., while the Medes and the Lydians were having a battle. "The day suddenly became night." Thales of Miletus, the first important Greek astronomer, is credited with predicting it. He may have based his prediction on an eclipse of May 17, 603 B.C. At any rate, the eclipse of 585 B.C. proved very useful. It caused the Lydians and Medes to cease fighting and make peace after five years of war.

While early observers for several millennia were awestruck by eclipses, their interest in systematic observations of the phenomena accompanying them seems indifferent. Apparently the corona was known to early Greek observers, Plutarch and Philostratus, but not until the time of Kepler, who witnessed the eclipse of 1605, was it recognized as something of importance. It was still another century before the red flames, now called prominences, came to be noted systematically with the European eclipse of May 12, 1706, which was the beginning of really scientific accounts of total solar eclipses. Totality lasted four minutes, with the corona visible as well as Venus, Mercury and Saturn and bright stars. Seen from the heights of the Swiss mountains, the stars were as thickly strewn as at full moon. At Montpellier the eclipse observers were greatly intrigued with animal behavior: "The bats flew about as at dusk. The fowls and pigeons betook themselves in great haste to their resting-places. The little birds which sung in cages were silent and put their heads under their wings. The animals which were at labour stood still."

The first eclipse to be well observed in the British Colonies of America was that of June 24, 1778. The observer was David Rittenhouse, whose name is so strongly linked to early intellectual development in the United States. Rittenhouse held various positions of importance, including that of the presidency of the American Philosophical Society, following Benjamin Franklin and preceding Thomas Jefferson.

Two years later the first American solar eclipse expedition went out from Harvard College under Professor Williams to study the eclipse of October 27, 1780, in Penobscot Bay. The eclipse was not total for these observers. Still, they described the ends of the crescent breaking off in fine drops of light. This is the phenomenon of Baily's beads, which were drawn to the attention of the world by Francis Baily many years later, at the eclipse of 1836. Other early total eclipse tracks fell on portions of America on June 16, 1806, in New York and November 30, 1834, in Georgia and South Carolina.

Baily's beads are caused by the rugged, mountainous edge of the moon, which seems to break up the rim of sunlight into droplets. Francis Baily, a stockbroker by profession, is an example of an amateur who has made a major contribution to astronomy. At an eclipse in 1842, conditions were so favorable for viewing them, as well as the brilliant corona and three large prominences, that it drew a tremendous burst of applause from the watchers.

A few years later photography began to be used in the observation of eclipses. During the eclipse of 1851, a daguerreotype recording the corona was taken in Germany with a special telescope known as the Königsberg heliometer. On July 18, 1860, when the track of totality crossed Spain, Warren de la Rue and Father Secchi of Rome obtained photographs revealing that the flames and corona were part of the sun, with the moon passing across them from minute to minute. Then eight years later, at the eclipse of August 18, 1868, in India and Malaya, a spectroscope was used on the prominences and proved that they were composed of hydrogen gas. And with the spectroscope came ways of observing the sun without an eclipse, by sorting out light of different wavelengths. This eclipse was memorable because Sir Norman Lockyer observed one of the Fraunhofer lines that was subsequently identified as due to a new gaseous element, and appropriately named helium, from Greek *helios* sun, after the body where it was first detected.

The enthusiasm for eclipse-chasing generated in the 19th century is still in evidence, and though we may understand the component parts of the occurrence better than did our predecessors a century ago, the feeling of awe is still with us.

Individual eclipses differ in their details, but the general phenomena are similar. More than an hour before the precious moments of totality, the dark disk of the moon starts to nibble away at the western edge of the sun. Larger and larger bites are taken out of the sun. This partial phase of the eclipse is the time when people can seriously hurt their eyes if they look fixedly at the sun. The surface brightness of the sun is not reduced as the moon passes over it — only the sun's total light is reduced. And it is the intense brightness of the surface that can destroy the human eye. You can use very heavily exposed film or smoked glass if you exercise great caution to see that enough of the light is being cut out. Of course to look directly at the sun with binoculars or telescopes unless they are equipped with filters or solar eyepieces would cause terrible eye damage. No eye protection is needed during the few minutes of totality.

One of the safest ways to view a partial eclipse is by the projection method, — that is, a pinhole camera. Simply by punching a pinhole in a firm piece of cardboard and projecting the image on a smooth surface, you can safely watch the progress of the eclipse. You might even see a sunspot in this way, if there happens to be one on the surface large enough for naked-eye visibility.

A neat trick is to form such a pinhole by clenching your fist and projecting the sun's crescent through the resulting narrow hole. At some eclipses pinhole cameras are automatically provided for the viewer. This occurs if he is standing near a tree with leaves. The spaces between the leaves act as pinhole cameras and you can see the ground covered with little crescents. Without an eclipse the same leaves would be casting round images of the sun which, in their thick overlapping, disguise their true nature.

As the light becomes more heavily blocked off, the behavior of nearby animals and birds always attracts interest. Birds head for their roosts or nests and animals for the barn. At the eclipse of February 15, 1960, an expedition of boys from an English school, Marlborough College, at La Turbie, overlooking Monaco and Monte Carlo, made a recording showing the fall-off in songs and chirps just before and during totality. At the eclipse of July 10, 1972, I watched flocks of cormorants that were fishing in the Bay of Chaleur return to their nesting island near Grand Anse, N.B.

At some eclipses just before or after totality a rather weird effect is produced by the shadow bands, waves of undulating darkness sweeping over the earthly landscape. They can be particularly prominent when the ground is snow-covered. The same conditions that make the stars twinkle are thought to cause them. The tiny crescent of the sun is being refracted down by air layers of differing thickness and density. Shadow bands are exceedingly difficult to photograph because they are of fleeting nature and the darkened landscape requires long exposures.

Also just before and after totality, you will see Baily's beads. And at the instant before and the instant after totality comes the diamond ring effect, a favorite with photographers. The last small area of light coming directly from the sun's disk is squeezed out in a flash. Combined with the thin circle of light, coming mainly from the inner corona, this gives an effect remarkably like that of a diamond ring.

As the sky darkens, the bright planets and brightest stars begin to be visible. If you can bear to miss the last 10 minutes or so of the partial phase, with the interesting effects that accompany it, you can get your eyes dark-adapted by blindfolding yourself. This really pays, for when you get a signal that totality has started, you open your eyes to see a sight you can hardly believe, so magnificent as to defy description. Instead of a dazzling body without visible structure, the sun has become extremely beautiful and complex. Around the dark disk of the moon, the delicately colored pearly white corona, something like cottony tufts, may have streamers extending outward for several solar diameters. And at the brilliant narrow base of the inner corona you may see bright red prominences, glowing masses of hydrogen gas extending upward 100,000 miles or so from the sun's edge. Near sunspot minimum the corona has short polar streamers, and very long ones extending out from its equatorial regions. At sunspot maximum the corona is symmetrically shaped, like a dahlia.

The minutes during which you can enjoy this rare sight are numbered, so plan to make the most of it as rapidly as possible, with the things you want to do, such as picture-taking. It is estimated that if a human being in an average adult lifetime went to every total solar eclipse and they were all clear, he would have only about an hour and a half in all in which to see the corona! Nowadays this time can be lengthened substantially if at each eclipse he is in an aircraft flying at high speed in the moon's shadow, and thus in effect making the moon stand still over the face of the sun.

While you are watching the eclipse progress, and particularly during totality, you will feel the chill of night rapidly descending. At the 1970 eclipse a drop of 7°C. or more was measured in the track of totality.

And then in an instant, with a burst of light from the sun, the total eclipse is over. You may have had anywhere from a few seconds to a few minutes of totality.

A geometrical requirement for a total eclipse is that the moon must always be at the new phase, since that places it directly between the earth and the sun. The greatest width of the moon's shadow cone where it touches earth is 167 miles, but the width of the track of totality is usually considerably less than this, frequently around 100 miles. This narrow path sweeps across an arc many thousands of miles long, and is the region where the eclipse may be seen as total. On either side of the path of totality, a partial eclipse may be seen over an area about 2,200 miles in width. For eclipses in high latitudes or those occurring near sunrise or sunset, this width may extend to 3,000 miles. You have a good chance of seeing a partial eclipse of the sun from your own home. The closer a place is to the actual track of totality, the greater the proportion of the sun's diameter that is obscured. The line of demarcation can sometimes be determined with precision. It was well shown in New York City in the famous eclipse of January 24, 1925. The dividing line was at 96th Street. Persons above that street were treated to a total eclipse, while those below got only a partial!

If the conditions for a total eclipse of the sun are met except that the positions of the moon and earth in their orbits cause the apparent diameter of the moon to be too small to hide the disk of the sun completely, then an annular eclipse results instead of a total. The name is derived from the Latin *annulus*, ring of light. With a ring of strong light from the surface of the sun left shining, none of the delicate phenomena which accompany a total eclipse is visible. Actually, annular eclipses are almost as rare as total, and are seen by fewer people, because not many persons make an expedition to view one.

There is no doubt that during the present century more people have seen a total eclipse of the sun than ever before in the world's history. Several tracks have crossed over or near very populous areas. In addition, the advent of television has brought most of the phenomena of an eclipse (except for the drop in temperature!) to the armchairs of millions. For example, the color TV coverage of the eclipse of March 7, 1970, was outstanding and made millions of people appreciative of the beauty of a total eclipse. It helped also to make people see the fallacy of the frequently expressed sentiment, "Well, it was 90 percent eclipse where I was, practically as good as total."

We have been fortunate in North America to have in a half-century a number of fine tracks. These eclipses were:

1923 September 10: California (cloudy), Mexico clear.
1925 January 24: Minnesota to Connecticut and out to sea.
1932 August 31: Sweeping down from Hudson and

James Bay, across eastern Canada and Vermont to Maine.

1943 February 2: Extreme northwest of North America, ending at Arctic Circle in Yukon.

1945 July 9: Coming southwest to northeast across parts of western U.S. and central Canada.

1954 June 30: North-central U.S. and central and northeast Canada.

1959 October 2: Extreme eastern edge of central New England.

1963 July 20: From western to eastern edge of North America, from Alaska all across Canada to northeastern tip of U.S.

1970 March 7: Sweeping northeast across Mexico, southeastern edge of U.S. and extreme eastern Canada.

1972 July 10: Alaska, northern Canada to Atlantic provinces.

If you have seen a total solar eclipse in North America and are trying to remember where and when, this table may help you pin it down. The eclipses seen by the largest number of people, and most apt to be remembered because of good weather conditions, are those of 1925, 1932, 1963 and 1970.

On February 26, 1979, a fine solar eclipse track will cross the very northern regions of the United States and move northward into Canada and across Hudson Bay, Ungava, Newfoundland and Greenland. After that, according to Jean Meeus, Carl C. Grosjean and Willy Vanderleen of Belgium (their *Canon of Solar Eclipses* contains maps of all the tracks from 1898 to 2510 A.D.), there will be a regrettable lull of 38 years, with no total solar eclipse tracks across the continental United States and Canada. The next one there will be on August 21, 2017. The next in Canada is in the High Arctic, higher than about 68°N., on August 1, 2008. Then, on April 8, 2024, there will be a total track in southern Canada, sweeping up from Mexico, across the United States and going out to sea across Nova Scotia and Newfoundland.

Two annular eclipses will have long tracks over well-settled U.S. territory on May 30, 1984, and May 10, 1994.

Total and Annular Solar Eclipses

Date	Type	Duration (min.)	General Area
1976 Apr. 29	Annular	7	* Africa, Asia Minor, S.E. Asia
Oct. 23	Total	5	S. Africa, Indian Ocean, Australia
1977 Apr. 18	Annular	7	S. Africa
Oct. 12	Total	3	* Pacific Ocean
1979 Feb. 26	Total	3	N. America
Aug. 22	Annular	6	Antarctic
1980 Feb. 16	Total	4	Africa, India, S.E. Asia
Aug. 10	Annular	3	Pacific, S. America
1981 Feb. 4-5	Annular	1	Tasmania and South Pacific
July 31	Total	2	* Central Asia and Pacific Ocean
1983 June 11	Total	5	Indian Ocean, Indonesia, Pacific Ocean
Dec. 4	Annular	4	* Atlantic Ocean and Africa
1984 May 30	Annular	1	* Mexico and southeastern U.S.
Nov. 22-23	Total	2	New Guinea to eastern South Pacific Ocean
1985 Nov. 12	Total	2	South Pacific Ocean and Antarctica
1986 Oct. 3	Annular-Total	2 sec.	* Greenland and Iceland
1987 Mar. 29	Annular-Total	1	South America, South Atlantic, Africa
Sept. 23	Annular	4	Central Asia, China, Samoa Islands
1988 Mar. 17-18	Total	4	* Indian Ocean, Philippines, Gulf of Alaska
Sept. 11	Annular	7	Africa, Australia, South Pacific
1990 Jan. 26	Annular	2	Antarctica and South Atlantic Ocean
July 22	Total	3	Northern Europe, Asia to Pacific Ocean

*Partial phase visible in parts of North America, as are partial eclipses of July 20, 1982, and May 19, 1985.

From computations of Julena S. Duncombe, U.S. Naval Observatory.

As pointed out earlier, we do not have an eclipse of the sun at every new moon because the moon's path is inclined 5° to the ecliptic. This means that usually the new moon will pass above or below the sun in the sky. The key factor in making eclipses possible is the proximity of the sun to a node of the moon's orbit, a point where the moon crosses the ecliptic. The interval of time when the sun is near enough to a node for an eclipse to be possible is known as an eclipse season. Eclipse seasons last about one month and occur twice in an eclipse year of 347 days. (An eclipse year is the interval of successive returns of the sun to the same node of the moon's orbit.)

Each calendar year brings at least two eclipse seasons in each of which there must be an eclipse of the sun. The eclipse may be only partial, or total or annular always accompanied by a partial phase. If the moon is new at the start of such a season, it may be new again before the season ends, making two solar eclipses possible at an eclipse season. And since the eclipse year is 18 days shorter than the calendar year, when an eclipse season begins in early January we will have two and a half eclipse seasons in one calendar year. This makes five solar eclipses possible in one calendar year, two in each eclipse season and one in the next half-season.

People frequently ask if eclipses can occur in any month of the year. Obviously they can, but the 19 days' difference in length between the eclipse and calendar years means that eclipse seasons occur that much earlier on successive years. In 1970 there was an annular eclipse of the sun August 31, but by 1972 the eclipse season was coming enough earlier to bring the total eclipse of the sun July 10. And the remarkable long-lasting African eclipse came June 30, 1973.

The nodes of the moon's orbit regress (move westward) around the ecliptic, completing the circuit in a little less than 19 years. This is a contributing factor to the saros, an interval after which an eclipse repeats. The discovery of this more than two millennia ago permitted the early prediction of eclipses. Thales probably knew of it, for the prediction of the eclipse on May 28, 585 B.C. basing his prediction on an eclipse of May 17, 603 B.C. The interval is 18 years, 10 1/3 (or 11 1/3) days, depending on the number of leap years in the period. It turns out that 223 synodic months or 19 eclipse years or 239 anomalistic months (from perigee to perigee) are approximately equal in length. So that after this interval the relative positions of the sun, earth and moon repeat. However, the third of a day over means that the eclipse is shifted 120° west in longitude. After three such shifts it is back to its starting point, but is shifted a little north or south from its earlier position. There are about 70 eclipses in a saros which begin near one pole of the earth and gradually work across the earth and off at the other pole in 1,400 years. Of these about 45 are either total or annular.

The saros shows up not merely in the dates when eclipses occur but also in the number of minutes of totality. The eclipse of June 30, 1973, was in a saros of eclipses of near-maximum duration. Four millennia ago, totality could last as long as 7 minutes, 34 seconds. As the earth's orbit shifts slightly, the maximum now is 7 minutes, 31 seconds. In an interval of 8,000 years (3000 B.C. to 5000 A.D.) only 62 eclipses have totality lasting 7 minutes or longer — about 1 percent of the total and annular eclipses in the period. To achieve these long-lasting eclipses a very delicate balance in the positions of the earth, sun and moon is essential. They occur in June or July and in or near the tropics. The eclipse of June 20, 1955, with a duration of 7 minutes, 8 seconds, had the longest duration of any eclipse for 911 years. The next in that saros, June 30, 1973, with a duration of 7 minutes, 4 seconds, was the longest totality until June 25, 2150, with 7 minutes, 14 seconds. Topping the list for the future is the eclipse of July 16, 2186, with a duration of 7 minutes, 29 seconds, only two seconds short of the maximum.

But 20th-century scientists are not content to sit back and merely accept the length of totality that natural conditions provide. In all total eclipses now, some astronomers are flying above the clouds and racing along with the moon's shadow. The totality of the 1973 African eclipse was extended to an unprecedented 74 minutes by the use for the first time of a supersonic Concorde plane flying at an altitude of 53,000 feet. Hopefully our knowledge of the all-important sun will be similarly extended in the future by modern technology.

Eclipses of the Moon

An eclipse of the moon is a very different phenomenon from one
of the sun. It is caused by the passage of the moon through the
earth's shadow. When the sun is eclipsed, nothing happens to the
sun itself; just its light is blocked from reaching a portion of the
earth's surface. But when the moon is eclipsed, something does
happen to it because the moon is not a self-luminous body like
the sun. When sunlight is prevented from reaching it, the moon
is darkened and the surface temperature falls rapidly. Since the
moon can always be seen above one-half of the earth's surface,
any eclipse of the moon is thus visible from half the surface of the
earth. Furthermore, a lunar eclipse may last as long as $3\frac{1}{2}$ hours.
This means that the turning of the earth will render the eclipse
visible over substantially more than half the earth.

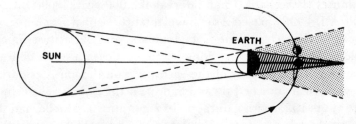

Figure 15. *A lunar eclipse occurs at full moon when the moon passes through
the earth's shadow.*

Observations of lunar eclipses were recorded very early. In fact
the Greek philosopher Aristotle in the 4th century B.C. using
such eclipses asserted that the earth was spherical. He gave as one
supporting reason that the shape of the shadow the earth casts on
the moon at eclipse is such as only a sphere could cast.
Unfortunately, his ideas were drowned out in the prevailing flat-
earth concepts held by the public for many centuries.

The geometrical circumstances that cause a lunar eclipse are
not so delicately balanced as in the case of a solar eclipse. The
earth carries constantly with it in space a conical shadow 859,000
miles long. Since the moon averages only 239,000 miles from us,

it can pass through a fairly wide part of the shadow, where it is as much as 2 2/3 times the moon's diameter.

Since to go into eclipse the moon must be directly behind the earth as seen from the sun, obviously an eclipse of the moon can occur only at full phase. Sometimes people say, "Isn't it fortunate that the eclipse of the moon occurred when the moon was full? It wouldn't have been nearly so interesting at some other phase." Actually, if an eclipse of the moon came at anything but the full phase, the astronomers of the world would be stunned.

Like solar eclipses, lunar eclipses can occur only during eclipse seasons. When the moon is only in the outer, lighter portions of the earth's shadow, the eclipse is barely noticeable and called penumbral. Umbral eclipses, where the moon gets into the deeper portions of the shadow, are very conspicuous. If all of the moon passes into the shadow (either umbra or penumbra) the eclipse is total. If a portion of the moon escapes the shadow, the eclipse is partial.

At each eclipse season a lunar eclipse must occur, but only one. It may be either penumbral or umbral. Each year we have two lunar eclipses, and if a third eclipse season comes along, then three. But they may be penumbral or umbral, partial or total. The maximum number of all eclipses in a year is seven. There may be five of the sun and two of the moon, or four of the sun and three of the moon.

From the geometry of the situation, you might expect that the moon would disappear during each eclipse. The reason that it does not is due to the earth's atmosphere. If the earth were only a hard solid body without atmosphere (and therefore without humans!) the moon *would* not be visible at all during total eclipse. The solid mass of the earth would stop the rays of the sun dead in their tracks before they reached the moon.

What actually happens is that the sunlight is refracted through the earth's atmosphere and some of it manages to reach the moon. The longer waves of light — the red rays — are the most facile in this respect, and the shorter blue rays are tossed aside. So even during the total phase of the eclipse the moon is visible, perhaps shining with an eerie coppery color. At times it resembles a colossal orange hung up in the sky.

When the umbral eclipse starts, a bite appears on the eastern edge of the moon, getting larger and larger as the eclipse progresses and the moon moves deeper and deeper into the earth's shadow from the west. Sometimes even at total eclipse the moon has enough light so that you can see the dim outlines of prominent features like the maria.

An intriguing aspect of total lunar eclipses is the very great range in brightness the moon exhibits from one eclipse to another. At some the moon is much fainter than average. On rare occasions it has disappeared completely, even when observers sought to find it with a telescope! Abnormal atmospheric conditions on earth are responsible. A feebly shining eclipsed moon may be the result of heavy clouds in the sections of the air through which the sunlight is trying to filter. And of course large amounts of volcanic dust diminish its transparency. In fact, the few records of complete disappearance or extreme faintness have mainly come from eclipses within a year or so of great volcanic eruptions.

Observers who have been treated to a disappearing moon could hardly believe their eyes, and understandably so. Three cases of total disappearance are recorded. Hevelius noted no trace of the moon in his telescope during the eclipse of April 25, 1642, the year after an eruption of Vesuvius, though the sky was covered with sparkling stars and the air was perfectly clear. A similar thing happened on May 18, 1761, according to Wargentin, when 11 minutes after the total phase began, the moon's body completely disappeared in his telescope, in a clear sky. This eclipse followed an eruption in Peteroa, Chile, in 1760. The last eclipse during which the moon disappeared completely in a telescope was that of June 10, 1816, observed in London and Dresden. The reason for the unusual state of the atmosphere then is not so obvious, but the summer of 1816 in England was one of the wettest summers in the century.

More common, but still rare, are eclipses in which the moon is so faint that it disappears with the naked eye but is still visible with a telescope.

A faint eclipse that is well known is that of October 4, 1884, following the eruption of Krakatoa, which brought colored suns and moons. On the evening of the eclipse in England there was a conspicuous return of the sunset afterglow, which had been common the previous winter. During the middle of totality nothing could be seen with the naked eye but a faint nebulous spot, though faint stars were visible above and below the moon. During the next lunar eclipse, March 30, 1885, an observer in Tasmania recorded that everything within the shadow was "lost in the dead slaty tint of the sky."

The most recent instances of extraordinarily faint eclipsed moons came after the explosion of Mount Agung volcano on Bali March 17, 1963. Just as with Krakatoa, though on a lesser scale, the volcanic dust was responsible for brilliantly colored sunsets around the world, and for the moon's being almost invisible in eclipse. The darkest eclipse was that of December 30, 1963. At that time I saw the moon turn to pale ashes and look like a faint piece of Milky Way clouds. Unless you looked for it in the sky, you would not notice it. The brightness of the moon dropped 4 magnitudes from the beginning of totality to mid-eclipse, when the moon had only as much light as a 4th-magnitude star.

At the next lunar eclipse June 24-25, 1964, the moon had the brightness of a 3rd-magnitude star. Its disk in eclipse was a dark uniform gray, with no detail or color. By the following eclipse, December 18, 1964, at mid-totality the moon was as bright as a zero-magnitude star, that is, it was 40 times as bright as one year before. The dust from the Mount Agung explosion was gradually settling around the world and atmospheric conditions were returning to normal.

The total eclipse of May 24-25, 1975, was a noticeably dark one, but from many reports, probably not quite as dark as that of December 30, 1963. However, in various regions such as southern Ontario, the effect of a light haze added to the darkness rendered the totally eclipsed moon at times invisible to the unaided eye.

At some eclipses the moon is much brighter than average. Such an eclipse was that of March 19, 1848, when the shaded surface had many times the brightness of an average eclipse. The features of the moon could be seen almost as well as on an ordinary dull moonlight night. Many persons observing it doubted that the moon was really in eclipse!

The explanation of abnormal brightening of the moon

involves the state of activity of the sun. Current research is under way to measure systematically the brightness of the moon in total eclipse. With weather satellites showing the global state of the earth's atmosphere every day, and solar observatories monitoring the state of activity on the sun's surface, eventually the causes that strongly alter the brightness of the eclipsed moon will be understood. But do not think because you have seen one lunar eclipse, you have seen them all. Some night you may get a real surprise!

Future Lunar Eclipses (Ephemeris Time)

Date	Type	Moon Enters Umbra	Total Eclipse Begins	Middle of Eclipse	Total Eclipse Ends	Moon Leaves Umbra
1976 May 13	Partial	19^{h}16.5^m		19^{h}55.1^m		20^{h}33.7^m
1977 Apr. 4	Partial	03 31.0		04 19.0		05 07.2
1978 Mar. 24	Total	14 33.6	15^{h}37.6^m	16 23.2	17^{h}08.9^m	18 12.8
Sept. 16	Total	17 21.0	18 25.2	19 05.0	19 44.7	20 48.9
1979 Mar. 13	Partial	19 29.7		21 08.8		22 48.1
Sept. 6	Total	09 18.7	10 32.1	10 55.0	11 17.9	12 31.3
1981 Jul. 17	Partial	03 25.6		04 47.6		06 09.8
1982 Jan. 9	Total	18 14.4	19 17.5	19 56.7	20 35.9	21 38.9
1982 Jul. 6	Total	05 33.7	06 38.6	07 31.8	08 24.9	09 29.9
Dec. 30	Total	09 51.3	10 59.0	11 29.6	12 00.1	13 07.9
1983 Jun. 25	Partial	07 15.3		08 23.2		09 30.9
1985 May 4	Total	18 17.5	19 22.9	19 57.3	20 31.6	21 37.1
Oct. 28	Total	15 55.5	17 20.7	17 43.3	18 05.8	19 31.0
1986 Apr. 24	Total	11 03.8	12 11.2	12 43.5	13 15.8	14 23.2
Oct. 17	Total	17 30.1	18 41.6	19 18.9	19 56.1	21 07.6
1988 Mar. 3	Partial	16 06.7		16 13.7		16 20.7
Aug. 27	Partial	10 08.4		11 05.5		12 02.6
1989 Feb. 20	Total	13 44.4	14 56.6	15 36.3	16 15.9	17 28.1
Aug. 17	Total	01 21.6	02 20.9	03 09.1	03 57.4	04 56.6
1990 Feb. 9	Total	17 29.5	18 50.2	19 12.0	19 33.8	20 54.4
Aug. 6	Partial	12 45.1		14 13.2		15 41.5

Note: Eclipses penumbral only: 1976 — April 14, November 17; 1977 — September 27, October 26; 1980 — March 1, July 27, August 26; 1981 — January 20; 1983 — December 19-20; 1984 — May 15, November 8; 1987 — April 14.

Courtesy of Dr. R. L. Duncombe, U.S. Naval Observatory.

Plate 1. Sundog at Churchill, Manitoba, near rocket launching tower.

Plate 2. Loading the Ahnighito Meteorite in Greenland, 1897.

Plate 3. New Quebec Crater, a meteorite hit in Ungava.

Plate 4. Earthshine on the moon — the old moon in the new moon's arms.

Plate 5. The full moon photographed with 100-inch telescope on Mount Wilson.

Plate 6. Constellations in the mail. Upper, the Big Dipper. Lower, the Southern Cross (Australia).

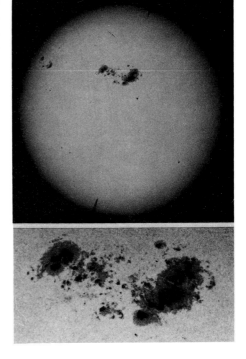

Plate 7. The great sunspot of April 7, 1947, photographed at Mount Wilson Observatory.

Plate 8. Total solar eclipse of July 10, 1972, at Canoe Cove, Nova Scotia.

Plate 9. Comet Ikeya-Seki and zodiacal light at Las Cruces, New Mexico, October 31, 1965.

MARS

JUPITER

SATURN

PLUTO

Plate 10. Four planets photographed with Hale Observatories telescopes.

Plate 11. A surprising sinuous valley 250 miles long on Mars. Mariner 9, February 2, 1972. (Photo center 29°S. Lat. 40°W. Long.)

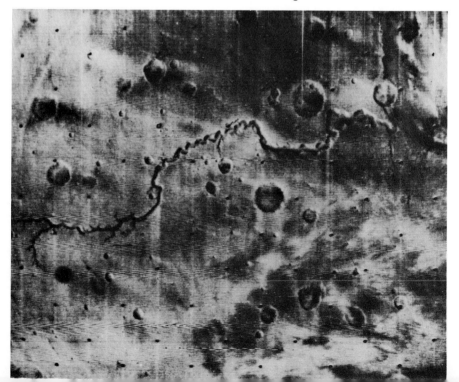

Eclipses Beyond the Solar System

Eclipses of the sun and moon represent but a tiny fraction of the eclipses that can be seen from earth even with the unaided eye. From a universe-wide view our eclipses of sun and moon are rather insignificant compared with some eclipses. These involve pairs of huge stars revolving around a common center of gravity. At certain times the larger star may completely block the light of the smaller from reaching a large area of space. These eclipses, staged by "Performing Stars," are discussed in Chapter 9.

★

★

★
SIX

The Planets

Star that bringest home the bee,
And sett'st the weary labourer free!
If any star shed peace, 'tis thou
That send'st it from above.

> T. CAMPBELL, "To the Evening Star"

Our seven-day week had its beginnings in western Asia, whence it spread to Europe and North Africa. The most logical explanation for a week of seven days is that it was in honor of the seven wanderers among the stars, the sun and moon and the five naked-eye planets. (The word "planet" means wanderer.) The names we use for the days of the week are the Saxon substitutes for those of the early Greek gods whose names were given to these wandering objects. Sunday is the sun's day (Sol); Monday, the moon's day (Luna); Tuesday, Tiw's day (Mars); Wednesday, Woden's day (Mercury); Thursday, Thor's day (Jove); Friday, Frigg's day (Venus); and Saturday, Seterne's day (Saturn).

The wandering feature of the bright planets drew the attention of the early sky-watchers and caused them, quite rightly, to attach considerable importance to these bodies. Ironically, it is this same feature that seems to prevent busy modern man from recognizing them when he sees them in the heavens! Because the planets move among the stars, they appear in different constellations from year to year, or even from month to month. And of course they share in the diurnal and annual shift in the sky as well. While you may early learn to recognize the North Star, Polaris, and you know just where to look for it any clear night in the year, you can never look for one of the bright planets in the same spot in the heavens all year long. The fact that planets usually don't twinkle will help you to distinguish them from stars.

All the five bright planets are very easy to see with the unaided eye. But without binoculars or a telescope you can see no distinctive markings on the surface of any of them to help you decide which one it is. You may make a guess at identification — Venus by its outstanding brilliance, for example. But a guide to the current positions of the planets is essential for informed stargazing. There are several good annual almanacs that tell where to find a planet on a given date, the *Old Farmer's Almanac,* or a more comprehensive one, *The Observer's Handbook* of the Royal Astronomical Society of Canada. Your enjoyment of the heavens will be greatly heightened by knowing which planet you are looking at, and by trying to visualize what conditions are like on it.

The planets fall logically into two groups from their positions with respect to earth. Those inside the earth's orbit, closer to the sun, are called the inferior planets. Those outside, farther from the sun, are known as superior planets. The first group consists of only two planets, Mercury and Venus. In the last century, astronomers thought they had detected a planet closer to the sun than Mercury. They named it Vulcan, a little prematurely as it turned out, because the existence of Vulcan is now disproved.

The second group consists of six known planets and is open-ended, for we may not yet have detected all the planets in the far reaches of the solar system. It also includes thousands of planetary fragments and tiny bodies up to 620 miles in diameter — the asteroids. One of these, Vesta, can be seen with the unaided eye under optimum conditions.

Although both groups of planets, like the earth, are revolving around the sun, the motions of inferior and superior planets as seen from the earth appear totally different. Those nearer the sun travel more rapidly than the earth, those more distant travel more slowly. Early in the 17th century the physical laws, three in number, under which the planets move were worked out by Johannes Kepler at Prague: (1) The orbit of each planet is an ellipse with the sun at one of the foci. (2) The line (radius vector) from the sun to the planet sweeps out equal areas in equal times. (3) And the squares of the periods of revolution of any two planets are in proportion to the cubes of their mean distances from the sun.

The Apron-String Planets

The two inferior planets Mercury and Venus are aptly described as tied to the sun's apron strings. There is a geometric limit to the number of degrees that can separate them from the sun. It is

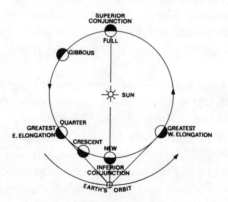

Figure 16. *The aspects and phases of an inferior planet.*

Solar System Data

Object	Physical Measurements						Orbit Around Sun				
	Equatorial Diameter Km.	Mass ⊕ = 1	Mean Density water =1	Surface Gravity ⊕ = 1	Rotation Period	Albedo	Mean Distance from Sun (a) A.U.	Mean Distance from Sun (a) millions of Km.	Period of Revolution Sidereal (P)	Period of Revolution Synodic days	Eccentricity (e)
Mercury	4,865	0.055	5.46	0.38	58d16h	0.056	0.387	57.9	88.0d.	116	.206
Venus	12,110	0.815	5.23	0.90	243d(retro.)	0.76	0.723	108.1	224.7	584	.007
Earth	12,756	1.000	5.52	1.00	23h56m04s	0.36	1.000	149.5	365.26	...	.017
Mars	6,788	0.107	3.93	0.38	24 37 23	0.16	1.524	227.8	687.0	780	.093
Jupiter	143,000	318.0	1.33	2.64	9 50 30	0.73	5.203	778.	11.86y.	399	.048
Saturn	121,000	95.2	0.69	1.13	10 14	0.76	9.539	1427.	29.46	378	.056
Uranus	47,000	14.6	1.56	1.07	10 49	0.93	19.18	2869.	84.01	370	.047
Neptune	50,900	17.3	1.54	1.08	16	0.62	30.06	4497.	164.8	367	.009
Pluto	5,500?	0.11	5?	0.6?	6d9h17m	0.14?	39.44	5900.	247.7	367	.250
Sun	1,392,000	332,958	1.41	27.9	25d.35d†						
Moon	3,476	0.0123	3.36	0.16	27d07h43m	0.067					

†Depending on latitude.

⊕ = earth

A.U. = astronomical unit, the earth's mean distance from the sun.

Adapted from *The Observer's Handbook.*

technically the angle subtended by the radius of the planet's orbit as seen from the earth. This limit is 28° for Mercury and 47° for Venus. Neither planet can ever be seen in the opposite part of the sky from the sun. However, Venus may still be above the horizon as much as four hours after sunset or four hours before sunrise.

The diagram on page 118 shows the positions (known as aspects) that these planets take up as they swing around the sun, as seen from earth. To simplify the diagram, the earth is held stationary, though of course it is continually moving too.

In a telescope these planets show phases just like the moon, because all their light is coming from the sun and each will have a dark, night side. Each aspect has a corresponding phase. For both Mercury and Venus the phases and aspects are the same. When either planet is on the opposite side of the sun its aspect is superior conjunction. Its phase is full because we are looking at the side on which the sun is shining. Its apparent diameter is then at its smallest. As the planet moves toward us, it increases in size and brightness, though the percentage of its illuminated disk that we can see steadily decreases. During all this part of its course the planet is evening star. When it comes to the aspect of greatest elongation east, and the phase corresponding to half-moon, it has reached its maximum angular distance from the sun. Then it seems to make a very swift retreat, with a diminishing crescent phase, and move back quickly into the region around inferior conjunction. There its phase is new, invisible, but it is lost in the sun's glare anyway. When it emerges from inferior conjunction it is a morning star, visible in the east before dawn. Quickly it draws farther and farther from the sun, to the aspect of greatest elongation west. Then it slowly sinks back toward the sun and superior conjunction. This entire cycle of aspects and phases is run through in one synodic period, which for Mercury is 116 days, and Venus 584 days.

One apparent violation of the rule that planets do not twinkle may come during the times when the phase of an inferior planet has dwindled down to a mere ghost of a crescent. At such times the crescent may be so narrow that the changes in direction of the light coming from it may make the disk appear to twinkle. When either of these two planets is close to the horizon and twinkly, this in an indication to the observer that they are in their very narrow crescent stages, even though you cannot see the phase with the unaided eye.

Numerical facts about the planets are listed in the table on page 119.

Mercury

Of the five planets visible to the naked eye, the one that is most rarely identified is Mercury, even though it is not the faintest. Mercury is so close to the sun that it darts back and forth with great rapidity. You have to look in the right part of the sky at the right hour during a few weeks each year. Even the great Copernicus stated that he had never seen Mercury properly. Perhaps that was because of the high latitude of his native Poland, and perhaps in those days people had not yet figured out the best time to hunt for it.

The Greeks had two names for this planet, Mercury as evening star, and Apollo as morning star, though Parmenides in the 6th century B.C. is said to be the first to declare that the evening and morning stars are the same.

Obviously the time to hunt for Mercury is near its greatest elongations. That maximum separation of 28° from the sun is rare — something like 20° is more usual. It is easiest to find Mercury when its path makes a large angle with the horizon, so that the planet is well above it. The best time to watch for Mercury as evening star is near that greatest eastern elongation which occurs nearest the vernal equinox. And conversely, as morning star near the greatest western elongation occurring near the autumnal equinox. At these times the ecliptic, whose path Mercury follows approximately, makes the largest angle with the horizon. Hence the angular separation of Mercury from the sun is upward and as far above the horizon as possible.

For a period of 10 days or even up to two weeks, centered on these elongations, Mercury is an easy object to see about 40 minutes after sunset as evening star, or a similar interval before sunrise as morning star. Its magnitude ranges from 0.5 to −1.5. It is brightest about a week before greatest eastern elongation, a week after greatest western elongation. Usually it is seen against the twilight or dawn sky and you are not much conscious of the constellations around it. But sometimes it is in Taurus near the conspicuous group of the Pleiades or the bright red star Aldebaran, which it resembles in color and brightness.

Since its synodic period is only 116 days (the shortest of any planet), Mercury takes less than one of our years to go through all its aspects and phases three times. Mercury is our third nearest neighbor among the planets. Common courtesy suggests we should recognize it when we see it!

Mercury is a small planet, with a diameter of 3,100 miles. Being the nearest planet to the sun (it can come within 28,550,000 miles), it is the swiftest moving and at perihelion moves as fast as 36 miles a second. This closeness to the sun, coupled with its small mass — less than 6/100 that of the earth — means that it could not long retain much atmosphere. For decades astronomers postulated that Mercury's period of rotation and revolution about the sun were equal at 88 days. Therefore, Mercury would keep the same side toward the sun, just as the moon keeps the same side toward the earth. It was a great surprise when radar observations with the 1,000-foot radio telescope at Arecibo, Puerto Rico, indicated its sidereal rotation period is only 58.6 days. (This corresponds to a solar day of Mercury of 176 earth solar days.) This period is exactly two-thirds of the sidereal period of revolution, 87.969 days. So Mercury does not keep the same side toward the sun.

Much of the material earlier written about Mercury concerning its permanently dark side, and speculations on its surface and atmosphere, is obsolete now. After centuries when even in the best telescope Mercury was only a fuzzy white crescent with gray irregularities, man has close-ups of its surface. And spectacular they are. Photos of Mercury at first glance can easily be mistaken for those of the moon. Radar observations with the 64-meter antenna at Goldstone, California, in the 1960's gave significant data on several planets, including evidence of craters on Mercury. But after March 29, 1974, when Mariner 10, given a "gravity assist" as it flew past Venus, passed 600 miles from Mercury, scientists knew exactly what 40 percent of Mercury is like.

Mercury consists of a heavy earth-style center (perhaps an iron core?) encased in a lightweight, moonlike surface, heavily cratered. An impressive feature is a multi-ringed basin 1,300 kilometers in diameter. It has a very thin atmosphere (less than one hundred billionth that of the earth) composed mainly of helium. A helium tail streams out in the sun's direction. It also has a weak magnetic field — a surprise for a planet whose rotation is so slow. Mariner 10 measured a temperature range of more than 600°C. — the largest in the solar system, from daytime readings over 400°C. to night at −173°C.

There are craters everywhere — old and young — some may have formed near its origin over 4 billion years ago. There are a few cliffs, but the mountain ranges and fissures seen on the moon are not present. A fine layer of dust, perhaps a few centimeters deep, seems to blanket the surface.

About 13 times a century, when Mercury is at an inferior conjunction, it crosses directly in front of the sun. Because of the small angular diameter of Mercury a telescope is necessary to see such a transit. In recent years some of the transits of Mercury have been televised and have been seen by more people than all previous transits in the history of the human race. The last was on November 10, 1973. The remaining transits this century will occur on November 13, 1988; November 6, 1993; November 15, 1999.

Venus

The planet Venus is the brightest object in the sky after the sun and moon. Its substantial size, its nearness and the high reflectivity of its cloud cover account for its great brilliance. It is very easy to recognize in the sky, especially if you follow its wanderings. It comes nearer to us than any other planet, about 25 million miles at its nearest. It is the earth's sister planet, with a diameter of 7,700 miles. However, Venus has some substantial differences from the earth.

Venus goes around the sun once in 225 days. In recent years its rotation on its axis has been shown to take 243 days. For centuries its heavy clouds defied astronomers to see its surface. On July 22, 1972, the Soviet probe Venera 8 landed there but sent back no pictures. However, other valuable data were obtained from it and from Mariner missions. A landing vehicle must withstand the terrific surface temperature 480°C (900°F), and pressures of 90 times that at the earth's surface, equivalent to 3,000 feet down in our ocean. At last, in 1975, actual photographs of the surface were transmitted to earth by the Soviet Venera 9, which landed October 22. This was followed three days later by Venera 10 which landed 1,375 miles away from the first. The illumination at the surface is much greater than that earlier predicted. The photographs even show shadows! The landscape is markedly different at the two 1975 landing sites. The first site has a mixture of smooth rounded stones and angular fragments a foot or more across. The second shows a smoother landscape and interesting "pancake" stones, interpreted by Soviet scientists as eroded mountain formation.

On February 5, 1974, Mariner 10 passed 3,600 miles from Venus after a 94-day voyage from the earth to the inner solar system. It took 3,000 pictures of Venus. The ultraviolet cloud tops are about 45 miles above the surface. The visible clouds consist of sulphuric acid droplets. There is a rapid drift of the atmosphere westward in a four-day retrograde rotation.

Though Venus goes once around the sun in 225 days (its sidereal period), it takes 584 days to line up with the sun in all its aspects as seen from earth (its synodic period). It is farthest away and faintest at superior conjunction, about 160 million miles. More than six months later it reaches greatest elongation east, where it is most conspicuous as evening star. At this time no one who looks at the clear western sky after sunset can miss it. It hangs low and brilliant. You can glimpse it through the canyons created by the tall skyscrapers of big cities like Toronto. Great cities pale out or hide the sight of many beautiful celestial objects, but fortunately they cannot annihilate bright Venus. Because of its outstanding brilliance, Venus undoubtedly is the unidentified object of most of the sonnets and poems to "the Evening Star." The Greeks had two names for this planet also: Phosphorus as morning star and Hesperus as evening.

At greatest eastern elongation Venus makes an apparent swift turn and takes less than five months to reach greatest western elongation. As it passes inferior conjunction in this interval it becomes a morning star. Early risers will see it before dawn in the east. Then it starts the trek back to superior conjunction.

Dates of greatest elongations of Venus for the rest of the century are given in the table below.

Greatest Elongations of Venus 1977 to 2000

Eastern Elongations	Western Elongations
1977 Jan. 24	1977 Jun 15
1978 Aug. 29	1979 Jan. 15
1980 Apr. 5	1980 Aug. 24
1981 Nov. 10	1982 Apr. 1
1983 June 16	1983 Nov. 4
1985 Jan. 22	1985 June 12
1986 Aug. 27	1987 Jan. 15
1988 Apr. 3	1988 Aug. 22
1989 Nov. 8	1990 Mar. 30
1991 June 13	1991 Nov. 2
1993 Jan. 19	1993 June 10
1994 Aug. 25	1995 Jan. 13
1996 Apr. 1	1996 Aug. 19
1997 Nov. 6	1998 Mar. 27
1999 June 11	1999 Oct. 31

Note: There is no greatest elongation of Venus in the year 2000.

M. B. B. Heath, in *The Journal of the British Astronomical Association.*

When Venus is farthest away we see it near the full phase, with most of its disk illuminated. When it is closest, at inferior conjunction, its phase is new with the dark side toward us. Where in its orbit will it be brightest for us? The famous English astronomer Edmund Halley solved this geometrical problem. He worked on it in 1716 when Venus was creating a stir in the city of London because many people noticed it in the daytime. Halley's computations show that Venus reaches its greatest brilliance, with magnitude −4.4, 36 days before or after inferior conjunction. At that time the planet has the phase of a five-day moon.

Venus is readily visible to the unaided eye in the daytime, even at high noon, especially near its times of greatest brilliancy. However, you need to know approximately where to look, so that your eyes do not tire searching a bright sky background. One way is to sight *along* (not through!) a telescope that has been set to the planet's position. Another is to note its position in the sky at sunset, and on succeeding days find it earlier and earlier. When you get it before sunset then you are seeing it by day.

When people tell of seeing stars in the daytime, Venus is the most likely object of their gaze, though some of the other planets can also qualify. Contrary to popular notion, you cannot see stars any better in the daytime by going down a deep well. This was proved experimentally some years ago by Dr. Allen Hynek of Northwestern University.

From time to time there are reports that someone has seen the crescent of Venus with the unaided eye. Most astronomers think this is so difficult as to be almost impossible. Professor C. A. Chant of the University of Toronto figured the difficulty in this way. When brightest, Venus has an angular diameter of about 40 seconds of arc. This is the angular width of a circle one inch in diameter and 430 feet away. The thickness of the crescent is about one-quarter the diameter of the disk. So cut out of a sheet of black paper a circle one inch in diameter, and then trim off a crescent a quarter inch in diameter at its widest. Now put it in front of a bright source of light 430 feet away, and you have the conditions similar to those necessary to see the crescent of Venus.

The phases of Venus became one of the first observations to disprove the Ptolemaic theory, which held sway for 1,500 years. Galileo first saw them in his early telescopes, and eagerly followed them. In the Ptolemaic theory it was impossible for Venus to show a gibbous phase. When in fact Galileo saw the phase turn to gibbous, he published his discovery in an elusive way common in those times. It was a Latin anagram, the letters to be suitably transposed later if time confirmed his discovery.

"Haec immatura a me iam frustra leguntur: o.y."

"These unripe things are read by me in vain."

After some months, when he was sure he was right, he transposed the letters to give the anagram its true meaning:

"Cynthiae figuras aemulatur Mater Amorum."

"The Mother of the Loves [Venus] imitates the phases of Cynthia [the moon]."

At the time this was the strongest observational support the Copernican system had yet received. Aristarchus of Samos (310-230 B.C.), had earlier put forward the heliocentric hypothesis, but his ideas were swept aside by the Ptolemaic hypothesis of sun and planets going around the earth.

Transits of Venus

There is one time when Venus is visible to the unaided eye at inferior conjunction. That is on the rare occasion when it passes directly in front of the sun — a transit. By rare, we mean *rare.* In recent centuries the transits of Venus have occurred in pairs eight years apart, with an interval of 122 years between the pairs. (Over a period of many centuries, this value shifts.) The last pair of transits was 1874 and 1882, commented on by many noted people such as Mark Twain and Stephen Leacock. The next pair will occur on June 8, 2004, and June 6, 2012. Because we have for years been near the middle of the interval between the pairs of transits, they are little talked of and relatively unknown. As we approach the time when most people can look forward to seeing this rare happening, they will be mentioned more frequently. The following pair will be on December 11, 2117, and December 8, 2125.

Attempts to observe the earliest predicted transits of Venus brought forth the most rigorous astronomical expeditions that will ever be experienced on earth. For hardship and severity they will be exceeded only by expeditions into space. Not until the

17th century was it possible to predict when transits of Venus would occur. The particular pair of transits that called forth the greatest human effort was that of the 18th century, in 1761 and 1769. Half a century earlier, Edmund Halley figured that the best way to determine the distance of the sun from the earth was to make accurate observations of the time and length of the passage as Venus crossed the disk of the sun. (This uses Kepler's Third Law.)

It was necessary that these transits be observed in many spots around the earth, and Halley exhorted the generation of astronomers after he was dead to "diligently apply themselves with all their might in making this observation."

Halley had figured out a list of strategic places where the transits should be observed. These included the north coast of Norway, the Shetland Isles, Madras, the western shores of Sumatra, Pondicherry on the west shore of the Bay of Bengal, Batavia (where the Dutch had a celebrated mart) and Port Nelson (now Churchill) on Hudson's Bay. The attempts of many eager scientists to carry out these plans led to astronomical voyages of the greatest length and difficulty ever undertaken on earth.

The transit of June 6, 1761, was observed by 176 persons at 117 stations in various sections of the world. Even though in colonial America most of man's energy had to be expended in establishing living and provender in the New World, nevertheless some enlightened citizens made expeditions for these transits. Professor John Winthrop of Harvard persuaded the governor to enlist the aid of the House of Representatives of the Province of Massachusetts. The sloop belonging to the Province took Winthrop, his assistants and his instruments to Newfoundland, where the skies were clear enough for observation. The sun rose with Venus on its disk, and Winthrop was able to determine five positions of the planet on the solar disk, and the moment of egress, that is, when the transit ended. His observations were given considerable weight.

Eight years later, in January, 1769, the American Philosophical Society appointed a committee of 13 to observe the transit. The observers formed three groups, one in Independence

Square, Philadelphia, under John Ewing; one at Norriton under David Rittenhouse; one at Cape Henlopen under Owen Biddle. Reports of this transit are given in the first volume of the *Transactions* of the American Philosophical Society.

All the way from England came two men, William Wales and Joseph Dymond, to observe at one of the posts signified by Halley — Port Nelson. In order to be on hand for the June, 1769, transit they left England 13 months before, on May 31, 1768, and wintered in that frigid territory. In the summer they were plagued by the insects common to the region, described by Wales:

"I found here three very troublesome insects. The first is the moschetto, too common in all parts of America, and too well known, to need describing here. The second is a very small flie, called — the sand-flie. These in a hot calm day are intolerably troublesome: there are continually millions of them about one's face and eyes, so that it is impossible either to speak, breathe, or look, without having one's mouth, nose, or eyes full of them . . . . The third insect is much like the large flesh-flie in England; but at least three times as large; these, from what part ever they fix their teeth, are sure to carry a piece away with them, an instance of which I have frequently seen and experienced."

When winter came, Wales was no longer bothered by insects: "In January 1769 the cold became intense. The little alarm clock would not go without an additional weight, and often not with that. A half pint of brandy in the open air had strong ice on top in five minutes. Sleep became impossible because of the crackling of the beams from the expansive power of the frost, which sounded like the guns on the house, which were three pounders."

All these hardships did not go unrewarded, however, for skies were clear for the transit on June 3 and Wales and Dymond were able to make useful observations.

Of all the adventures in the 18th-century transits of Venus, the most dramatic is that of the Abbé Le Gentil. His 11-year voyage to the Indian Ocean to observe both transits, in 1761 and 1769, is the longest-lasting astronomical expedition in history. And perhaps his experiences represent the ultimate in astronomical frustration. At the age of 35 he was sent by the French Academy of Sciences to observe the transit at Pondicherry, a French possession on the southeast coast of India near the tip. On March 26, 1760, he embarked on the ship *Berryer* of the East India Company, sailing around the Cape of Good Hope to the Isle de France (now Mauritius). He had to wait there, severely ill, for eight months before he could get another ship bound for India. But on May 24 as they neared Pondicherry the captain learned that it had fallen to the English, who were at war with the French. The captain made a hasty turn around back to the Isle de France. Le Gentil observed the transit from the unsteady deck where his observations could yield nothing of value.

Undaunted, Le Gentil decided to be sure of being on the spot for the next transit by waiting out the ensuing eight years in the region of the Indian Ocean. He put these years to good use, however, in a study of winds and monsoons, the astronomy of the Brahmins, and in mapping the east coast of Madagascar. More than a year before the transit, he arrived at Pondicherry on March 27, 1768, and established an observatory in an abandoned fort over 60,000 weight of gunpowder. Obviously, his travel experiences had not made him nervous.

Though June mornings were almost always fine in Pondicherry, when Le Gentil awoke at 2 a.m. on the day of the transit, June 3, 1769, he saw with dismay and astonishment that the sky was cloud covered. At several minutes before seven, the important moment when Venus was scheduled to leave the sun's disk, only a light whiteness in the sky indicated the sun's region.

Le Gentil speaks for many astronomers whose plans have been ruined by clouds when he wrote dejectedly: "I had gone more than ten thousand leagues; it seemed that I had crossed such a great expanse of seas, exiling myself from my native land, only to be the spectator of a fatal cloud which came to place itself before the sun at the precise moment of my observation, to carry off from me the fruits of my pains and of my fatigues. . . . I was unable to recover from my astonishment, I had difficulty in realizing that the transit of Venus was finally over . . . At length I was more than two weeks in a singular dejection and almost did not have the courage to take up my pen to continue my journal; and several times it fell from my hands, when the moment came to report to France the fate of my operations . . ."

Le Gentil's troubles did not end with the clouds on the day of the transit. He had difficulties getting back to France, which he reached on October 8, 1771, after 11 years, six months and 13 days of absence — but not to a hero's welcome. His heirs and creditors had not heard from him, believed him dead, and were about to divide his estate, whose funds had been somewhat dissipated. And the height of irony was that he had been superseded in the Academy of Sciences, on whose behalf he had undertaken the voyage. Eventually these difficulties were all overcome, Le Gentil married, and spent many happy years writing his memoirs and other papers until he died in October, 1792, at the age of 67. It is highly unlikely that anyone ever again will make such an effort to see a transit of Venus as did Le Gentil.

The Frigid Globes

Only three of the planets beyond the earth, Mars, Jupiter and Saturn, are sufficiently bright that a casual observer notices them. These superior planets assume very different positions from those of the inferior planets. There is no apron-string tie here. It is part of their courses to be completely opposite the sun in the sky. During such a time they are closest to earth.

Near this aspect of opposition, they rise about sunset and are above the heavens all night. A superior planet drops steadily westward with respect to the sun. After opposition it comes to the

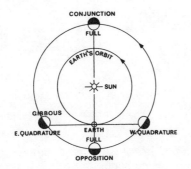

Figure 17. *The aspects of a superior planet.*

configuration known as quadrature east. (The term "quadrature" means that there is a right angle at the earth between the direction to the planet and to the sun.) Then, dropping farther west, the planet disappears in the twilight western sky as it nears the aspect of conjunction. A few weeks later it will emerge in the dawn sky as morning star, rising before the sun. Continuing to rise earlier and earlier, it reaches the aspect of quadrature west, when it rises near midnight. Continuing its earlier rising, it will once again come to opposition.

These are the motions of a superior planet with respect to the sun, and they determine when we will see it in the night sky. All this time the planet is moving steadily among the stars. Most of that motion is eastward, and very close to the ecliptic, which the orbits of the naked-eye planets tend to hug. When the planet nears opposition, however, the earth starts overtaking it so rapidly that the planet appears to be moving westward with respect to the stars — that is, backward. This is called retrograde motion. Its path forms what is known as a "loop of retrogression," which lasts some weeks or months (depending on the planet). The two ends of the loop, where the planet hovers briefly, are the "stationary points." This motion of the planet among the stars has been observed for several millennia. In order to account for this loop of retrogression the Ptolemaic theory of the solar system mistakenly resorted to a series of circles known as epicycles and deferents.

The motion of a superior planet among the stars is very easy to

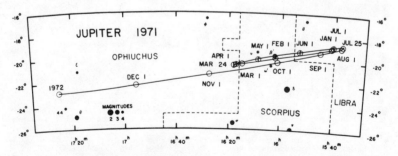

Figure 18. *The loop of retrogression of a superior planet. (From* The Observer's Handbook.*)*

follow. Simply note the position of the planet with respect to bright neighboring stars. Most almanacs give the stationary dates. Over a period of a few weeks you can watch it change direction.

In a telescope the superior planets do not show the same phases as the inferior ones do. We are always looking at approximately the side on which the sun is shining. The only phase effect is that the planet can be slightly gibbous near its quadratures. The effect is greatest for the nearest superior planet, Mars, and decreases with the more distant ones. But a spectacular view of Jupiter as a crescent was acquired by Pioneer 10.

Mars

Many people would vote for Mars as the most intriguing planet beyond the earth. A ruddy planet with a diameter of 4,200 miles, it comes nearer us than any other planet except Venus. With a sidereal period of 687 days, Mars has the most elliptical orbit (greatest eccentricity, .093) of any superior planet visible to the unaided eye. Thus its least distance from us, near opposition, can vary enormously, by about 28 million miles. The whole variation in distance of Mars from us is about 200 million miles.

Mars moves much faster among the stars than any of the other superior planets, taking only 780 days (two years and two months) to go completely around the sky and back to the sun, its synodic period. With the unaided eye you can follow Mars at all times except near conjunction. The ruddy color of Mars, which helps to identify it, has left its mark on the name of the red supergiant star in Scorpius, Antares, which means "Rival of Mars." When Mars gets into the region of the ecliptic near Antares, it is sometimes difficult to tell which is star and which is planet, but the twinkling of the star may help.

Closest approaches come when Mars reaches opposition near its own perihelion. The favourable oppositions come in our late summer or early fall every 15 or 17 years. At best, Mars comes within 35 million miles of earth. Then it outshines every planet

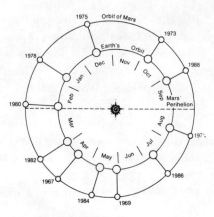

Figure 19. *Varying distance of Mars at oppositions. (By permission, from Baker and Fredrick,* Astronomy, *New York: D. Van Nostrand Co., 9th ed., 1971.)*

but Venus, and has an angular diameter of 25 seconds of arc. When seen near the horizon, it glows with such a red brilliance that it is often mistaken for a terrestrial light, or even a flying saucer.

At the opposition on August 10, 1971, with Mars nearest the earth on August 12, we had the conditions that produce virtually the minimum distance, 34,931,000 miles, and magnitude −2.6. At the 1973 opposition on October 24 the nearest approach was October 16, with a distance of 40,532,000 miles.

Although the Martian year is nearly twice as long as the earth's year, the Martian day is remarkably similar in length to ours — only 41 minutes longer. We see the surface of Mars through a very thin atmosphere, at least 90 percent of which is carbon dioxide. Its total atmospheric pressure is less than 1/100 that of earth's — similar to that at a height of 25 miles here. With only one-eleventh the mass of the earth, Mars is unable to hold a heavy atmosphere. Mars is a cold planet, though temperatures get well above freezing in the tropics by day.

The surface details have intrigued astronomers for decades, ever since Schiaparelli's Italian word *canali* became translated as "canals" (suggesting intelligent activity) rather than "channels," or natural features. The pictures from Mariner 9 show a varied

Oppositions of Mars

Year	Opposition	Nearest Earth	Millions of Miles	Magnitude
1975	Dec. 15	Dec. 9	52.6	−1.5
1978	Jan. 22	Jan. 19	60.7	−1.1
1980	Feb. 25	Feb. 26	63.0	−0.8
1982	Mar. 31	Apr. 5	59.0	−1.1
1984	May 11	May 19	49.4	−1.7
1986	July 10	July 16	37.5	−2.4
1988	Sept. 28	Sept. 22	36.5	−2.5
1990	Nov. 27	Nov. 20	48.0	−1.8
1993	Jan. 7	Jan. 3	58.2	−1.2
1995	Feb. 12	Feb. 11	62.8	−1.0
1997	Mar. 17	Mar. 20	61.3	−1.1
1999	Apr. 24	May 1	53.8	−1.4

Courtesy of Dr. R. L. Duncombe, U. S. Naval Observatory.

surface with high mountains and large basins, a desolate, crater-pocked desert similar to the moon. They show strings of craters, earlier called the *canali*. And they show surprising indications of earlier river valleys and flowing water. There is even a suggestion that perhaps as recently as 10,000 years ago Mars had water on its surface — a great climatic change brought about by its precessional cycle. The appearance and disappearance of different types of clouds make an intriguing and unpredictable feature. Heavy clouds sometimes impede observations at favorable oppositions. The yellow clouds are considered dust clouds, the white are thought to be crystalline, like cirrus on earth.

Polar caps form an outstanding feature, composed of ices of carbon dioxide or water, with an icy fog hanging over them. The northern cap extends at maximum to a radius of 30° and never disappears, but the southern cap does vanish for some weeks in summer.

Mars has two tiny moons. Phobos, dubbed "the potato moon" because its photographs resemble a potato, is 8.4 by 6.0 miles in size and 5,700 miles from the planet. Deimos, 4.7 by 3.4 miles in size, is 15,500 miles away. Mariner 9 took 36 photographs of Phobos, covering 80 percent of its surface. These are sufficient for an atlas with named features, including the largest crater, Stickney, 5 miles in diameter (one-third the size of Phobos!), and a curious long ridge, Kepler Ridge. This feature indicates that Phobos has the structural strength of rock. The heavily cratered surfaces of the moons show their age to be billions of years. Mariner 9 could photograph only the side of Deimos facing Mars, and the coverage of that moon is less complete.

The discovery of these moons is a fascinating astronomical puzzle. In his *Gulliver's Travels,* published in 1726, the learned Jonathan Swift, Dean of St. Patrick's Cathedral, Dublin, postulated their existence as well as their approximate periods of revolution and distances from Mars. This was a century and a half before our earthly telescopes had the capacity to reveal them! They were discovered in 1877 by Asaph Hall testing a powerful new telescope for the U. S. Naval Observatory.

Jupiter

While the sun possesses most of the mass of the solar system, it is the great outer planets that have in their orbital momenta, 98 percent of all the momentum in the solar system. Any theory of the origin and development of the solar system must explain this fact.

Most of the time Jupiter is the fourth brightest object in the sky, surpassed only by the sun, moon and Venus. At its brightest, around the time of its nearest oppositions, it is of magnitude −2.5. At every opposition it is as bright as −2.1. Only for rare and short periods is Mars brighter than this. Near conjunction, at its greatest distance from us, it may be as faint as −1.4, when the brightest star in the sky, Sirius, also excels it.

Another circumstance also helps to make Jupiter one of the planets you can recognize most easily. When we get as far out in space as Jupiter, we are coming to regions where objects, relatively speaking, stay put. While Mars is circling the heavens every two years, Jupiter is moving in a ponderous manner, taking 12 years to go once around the heavens. This makes a very convenient circumstance because there are 12 constellations of the zodiac. Accordingly, Jupiter stays for about one year in one of them. If you learn its position one year, the next year just move your glance one zodiacal constellation to the east and there it is!

With a diameter of 88,000 miles Jupiter is the largest planet. Precise measures from spacecraft now give the exact dimensions: the equatorial radius is 44,368 miles, and the polar 41,516 miles. These figures show a considerable flattening at the poles — oblateness — caused by rotation. Jupiter is also the most massive, several times more massive than all the outer planets combined. Indeed, of all the mass of the solar system which is not in our sun, more than half is in the planet Jupiter. Some of the lightweight bodies like asteroids and comets come under the strong gravitational control of this planet, and move in orbits that are tied to Jupiter.

Although it is easy to see the planet with the unaided eye and follow it around its course, a telescope is necessary to reveal details. It is the top of a heavy atmosphere we see, by contrast with Mars, where the actual surface is visible. It is even uncertain as yet whether Jupiter has a solid surface down underneath. Jupiter has the shortest rotation period of any planet, only nine hours, 50 minutes. This rapid whirling influences the structure of its atmosphere. The features that can be seen are of a semipermanent nature, some changing slowly over months, years or decades. Always present are broad equatorial belts. Over 300 years much of the time the most prominent feature has been the Great Red Spot. An oval spot, 25,000 miles long, particularly prominent a century ago, is still visible.

For centuries astronomers have tried to understand the planet. Now, with space-age Pioneer 10 and 11, the understanding is coming by leaps and bounds. In December, 1973, Pioneer 10 passed within 81,000 miles of Jupiter, taking as many pictures and measures as possible. At that time it was so far distant that the signals took 46 minutes to reach earth, traveling with the speed of light. Then it headed out beyond the solar system, bearing a plaque which eons hence may reach some distant civilization.

In December, 1974, Pioneer 11 came about 26,600 miles from the cloud top on a polar orbit, carefully calculated to propel the spacecraft through the dangerous radiation zones as rapidly as possible. Pioneer 10 had shown that the strength of the charged particles was 1,000 times that which would kill a man. A flood of 150 million protons could be expected to hit every square centimeter of the spacecraft every second! These did damage the immediate readings on Pioneer 11, but did not destroy the spacecraft nor impede it in the long run. Jupiter's strong magnetic field associated with the charged particles is one of the great space-age discoveries. It is like a great flabby bag that can stretch out over more than 9 million miles, but is sometimes pushed in by the solar wind to only half that dimension.

The colors of Jupiter seen by the Pioneers are brilliant reds, oranges, blues, yellows and whites. The equatorial white stripes on high resolution dissolve into smaller patches. Near the poles are deep indigo atmospheres with bubbly red circular patches in which one can see a little deeper down into the planet, perhaps 30 miles. The famous Great Red Spot is a noticeable feature in the southern hemisphere. Pioneer 10 photos seem to show it casting a shadow, apparently rising above the level of the surrounding clouds. It may be the upper-atmosphere result of some permanent feature deeper down in the planet. There is marked change between Pioneer 10 and Pioneer 11 photos of the Spot, which show the northern edge breaking up with large, dark patches of lower and warmer clouds.

Recent measurements reveal that Jupiter is not the frozen globe it was formerly thought, but is sending off into space about two and a half times as much heat as it receives from the sun. The source of this heat is not yet known. The picture of the planet now is a turbulent one — a heavy atmosphere with raging winds, strong heat flow as it spins swiftly and strong magnetic field. The atmosphere consists of 82 percent hydrogen, 17 percent helium and 1 percent other elements. It is known to be rich in compounds like ammonia. Carl Sagan suggests that this could be a warm, favorable brew for the simpler forms of life. The heavy atmospheric pressures and surface gravity make Jupiter an inhospitable place for humans, however.

About 6,000 miles down the atmosphere changes from gas to liquid hydrogen. The temperature and pressure both rise rapidly with depth. Perhaps the center is a small rocky core with a suggested temperature of 54,000°F. The pressure would be millions of earth atmospheres.

Another modern riddle of Jupiter is the source of its strong radio noise. When first picked up on radio telescopes, the bursts were so intense that they were thought to be terrestrial interference! The unknown source is producing energy equivalent to megaton hydrogen bombs going off at the rate of one a second.

Jupiter dominates the solar system in still another way. It has the largest number of moons of any planet, 13 in all, and two are larger than our moon. The 13th, named Amalthea, was discovered only in 1974, by Caltech astronomer Charles Kowal — the first discovery of a solar system moon since that of Jupiter XII in 1951. A probable fourteenth moon was discovered by Kowal in September, 1975.

The four large moons of Jupiter were discovered in the early 17th century by Galileo, and independently by Simon Marius, as soon as the first telescopes were turned to Jupiter. They are known as the Galilean satellites. Ganymede, the largest, with a diameter of 3,120 miles, is as big as the planet Mercury. Callisto is larger than our moon, and remarkable photographs from Pioneer 11 showed a south polar cap, which might consist of frozen water. Io is a little smaller, and one of the most remarkable of the moons. The strength of the radio signals we receive from Jupiter depends on the position of Io in its orbit. Furthermore, Pioneer 10 showed that Io has an atmosphere, only the second solar system moon shown to have one. Europa is the smallest of the Galilean satellites at 1,790 miles. Densities indicate the inner two may be largely rock and the outer two, Ganymede and Callisto, composed partly of ice or frozen carbon dioxide.

The four Galilean satellites are actually all bright enough to be seen with the naked eye, about the 5th magnitude. The catch is that they are so close to Jupiter, which averages about

800 times as bright. None of these moons can ever be more than 11 minutes of arc from it. Nevertheless, there are recorded, substantiated instances by especially keen-sighted persons who have seen one or more of these moons with the naked eye.

One well-documented claim is that of Lieutenant Elliot Brownlow of the Bengal Engineers taking part in the Trigonometrical Survey of India in the middle of the last century. Described as having exceptional vision, he was able with the unaided eye to make a sketch of the positions of the moons of Jupiter which was immediately verified by another observer using a telescope.

This is a real challenge for people with exceptional acuity of vision. It is by no means a trivial point. For if a person can really see the moons without a telescope, then the moons may very well have actually been discovered centuries or millennia ago, though probably the observers did not then realize what they were seeing.

These moons are one of the most interesting sights for a small telescope. They stage their own shows with disappearances behind Jupiter and into its shadow, and transits of themselves or their shadows in front of the ball of the planet. Various annual almanacs such as *The Observer's Handbook* give the satellite positions each night that Jupiter is well placed for observation. You can identify the satellites in that way.

These Galilean satellites have made a most important contribution to the history of science. Through the timing of their eclipses and occultations, the Danish astronomer Ole Roemer discovered that light travels with a finite velocity. Roemer found that when Jupiter was on the other side of the sun from the earth, the eclipses were later than their scheduled times. He reasoned that the extra time was needed for light to travel the added distance. It takes light eight minutes to go from the sun to the earth.

The other nine moons of Jupiter are faint and small, less than 100 miles in diameter. Jupiter V is very close to the planet, only about half as far as our moon from us. The others are all far away — millions of miles out. Jupiter XIII is about 7.7 million miles from Jupiter, and takes 210 days to go around it once. Its orbit and distance are similar to those of Jupiter VI, VII and X.

Saturn

Beyond Jupiter and almost twice as far away from us lies the most distant planet known to the ancients — Saturn. It is so far away that our earth, as seen from Saturn, never gets more than 6° away from the sun. If there were human beings on Saturn (highly unlikely) there would be a good chance that they have not yet discovered the earth. Hopefully, in a few years we will learn more about Saturn than ever before. Pioneer 11 has been given a corkscrew flip as it left Jupiter and, after traversing 1½ billion miles, will reach Saturn in 1979.

Saturn is the second largest planet, with a diameter of 75,100 miles and 95 times the earth's mass. It has the lowest density of any planet, only 0.7 that of water.

Like Jupiter, Saturn has colored but less prominent belts. The finest characteristic of Saturn is its magnificent ring system. This is not visible to the unaided eye and was not discovered until the telescope was invented. Even after Saturn had been studied by several observers with the first crude telescopes, several decades elapsed before the observations were correctly interpreted. In the first telescopes the ring system bulged out from the planet, looking like a moon on either side. Then these two supposed moons bewildered observers by alternately appearing and disappearing in the course of years. Actually, Saturn's rings show phases as their tilt toward the sun and earth changes in the 30-year period that Saturn takes to go around the sun. This is comparable to the change of tilt of our earth toward the sun during the year, a change that causes our seasons. Once in 30 years the northern face of the rings of Saturn opens its maximum amount of 27°, and once in the same period the southern face does likewise. Twice in 30 years the earth passes through the plane of these thin rings and for a brief period of time they may completely disappear. In 1966 we had the last closing down of the rings. The maximum opening of the southern face of the rings occurred in 1974.

The brightness of Saturn in the sky depends both on its distance from earth and on the tilt of the rings. The open rings reflect 2.7 times as much light as the ball of the planet itself. It is brightest, magnitude −0.4, when the rings are widest open and

Saturn comes to opposition near its perihelion. When the rings are closed it gets no brighter than magnitude +0.9. At its faintest, near conjunction, it is only of the 2nd magnitude.

The rings of Saturn are 171,000 miles from tip to tip. The thickness may be inches or miles. There are three rings with a conspicuous dark gap known as Cassini's division between the two bright ones. They are not rotating like a wheel in which the parts most distant from the center spin most rapidly. Instead, they obey Kepler's laws of motion with the particles nearest the center traveling fastest. The exact size and composition of the particles or small bodies that make up the rings is not well determined. Radar observations have surprised astronomers by showing amazingly high reflectivity in the rings. Whereas compared with the reflectivity of a silver sphere the moon reflects 5 percent and Mars 8 percent, the rings of Saturn send back 60 percent. And this despite the fact that the material in the rings is so loosely packed that you can sometimes see stars through it! Jagged hunks of material several feet across or spheres of ice a few inches in diameter could give this effect.

Saturn is the farthest planet out that can be well reached by radar from earth. At its nearest, 8.4 astronomical units, the signal takes $2\frac{1}{4}$ hours to reach Saturn and return. This means the beaming can continue for that interval and then the scientists listen for the next $2\frac{1}{4}$ hours.

Saturn has 10 moons, none visible to the naked eye. Titan, the largest, is easily visible in a small telescope. Titan is bigger than our moon, with a diameter of 2,980 miles, and was the first moon in the solar system on which an atmosphere was found. It is never brighter than the 8th magnitude. The tenth moon, Janus, was discovered only in 1966 by A. Dollfus at the last closing down of the rings. Less than 300 miles in diameter, it is the closest of all the moons to Saturn, only 100,000 miles from it.

Uranus

The outer boundary of the known solar system was more than doubled in the 18th century by the discovery of the planet

Uranus. "Uranus was the first planet that needed to be discovered," according to one of my students writing an exam.

The fourth largest planet, with a diameter of 29,000 miles, Uranus is a little brighter than the 6th magnitude at opposition. Under optimum conditions it can be seen with the unaided eye. It is easiest to locate first with binoculars; then try to recognize it with the unaided eye. Almanacs annually give maps of its position among the stars. Since it takes 84 years to go once around the sun, it stays in the same region among the stars for quite a while. At 19 times the earth's distance from the sun, Uranus is so far out that even a good-sized telescope does not reveal much about its greenish disk. However, the first evidence for molecular hydrogen in the atmosphere of a major planet was obtained for Uranus by G. Herzberg. A curious feature of the planet is that it rotates in 10¾ hours with its equator nearly perpendicular to its orbit, with an inclination of 97°. Uranus has five moons, all faint and visible only in a good-sized telescope. Their orbits also make an interesting up-on-end pattern.

The discovery of Uranus is one of the dramatic stories of astronomy. The famous astronomer William Herschel in England had constructed the finest telescopes ever made, and was systematically conducting a search of the heavens. On March 13, 1781, he discovered a strange object which he described in his journal, "In the quartile near ζ Tauri, the lowest of two is a curious either nebulous star or perhaps a comet."

On succeeding nights he observed that it moved among the stars. Its path was followed at other observatories in Europe, and it was considered to be a comet. But after some weeks the Russian astronomer Anders John Lexell, visiting in England at the time, showed that the new body was moving in a circular orbit at 18 times the earth's distance from the sun. The object then was not a comet but a new planet! No one had ever discovered a planet before. The famous lines of Keats, "Then felt I like some watcher of the skies when a new planet swims into his ken," hardly apply to the first planet to be discovered.

The new planet had a profusion of names. Herschel named it Georgium Sidus for his patron, George III. The French astronomer Lalande named it Herschel, toward whom the feelings on the continent were warmer than for the King of England. The German astronomer Bode named it Uranus in keeping with the mythological names of the solar system. After many years had passed, this name was adopted.

An interesting sidelight is that a search of early records showed that Uranus had been seen and recorded 20 times before its discovery. Even as recently as 1968 still another early observation was unearthed. But the other observers had not bothered to follow through on the identification! Sky watchers less careful than Herschel had missed being the discoverers of a new planet.

Neptune and Pluto

Beyond Uranus are the planets Neptune and Pluto, invisible to the unaided eye. And their discovery is a triumph of astronomy of the invisible. For decades after its discovery, Uranus kept running off its predicted course by a minute or two of arc. Finally two mathematical astronomers, John Couch Adams in England and Urbain Jean Joseph Leverrier in France, were convinced that the displacement of Uranus was caused by an unknown planet out beyond it. Independently they did the long calculations necessary to give the position and mass of the unknown body. Unfortunately, Adams' predictions were not followed through promptly with telescopic observations. Leverrier's were sent to Johann Galle of the Berlin Observatory who turned the big telescope to the search that very night. And there, within a degree of Leverrier's predicted position was the new planet!

Neptune, 31,650 miles in diameter, takes 165 years to go around the sun. Of the 8th magnitude, it is visible in good binoculars. It has two moons, of which Triton is larger than our moon.

The only planet to be discovered so far in the 20th century is Pluto. And its discovery is more controversial than that of Neptune. Early in the 20th century two astronomers from Boston, Percival Lowell, and William Pickering, became convinced that Neptune was straying from its path by the pull of a planet out beyond it. At his observatory in Mandeville, Jamaica,

Pickering worked to determine a position for the Trans-Neptunian planet. Lowell did likewise at his observatory, the Lowell in Flagstaff, Arizona. The search continued for years without result.

Then suddenly, after both Lowell and Pickering were dead, a young astronomer, Clyde Tombaugh, detected the image of the Trans-Neptunian planet on plates taken at the Lowell Observatory in a systematic hunt with a new telescope. The discovery was announced to the world March 13, 1930. This date is Percival Lowell's birthday as well as the date of the discovery of Uranus.

The planet had been detected some weeks earlier, but the observatory kept it secret until enough observations were acquired to prove that the object was indeed moving in an orbit out beyond Neptune. (Over a short interval an asteroid near the stationary point could be mistaken for a more distant planet.)

But the discovery was a great surprise. Instead of the planet astronomers were looking for, of about the 12th magnitude, Pluto turned out to be unexpectedly faint — of the 15th magnitude. It is a small planet, only 3,500 miles in diameter, rotating slowly in 6 days, 9 hours, 17 minutes. Pluto has the density of the inner planets, not the outer. And its mass is much less than expected — only 11 percent that of the earth. In fact, its mass seems to be too small to have pulled the two outermost planets off their courses. Hence its existence could not have been predicted. This leads to a most curious enigma.

Pluto itself is different from the other planets. It is the only one that crosses inside the orbit of another planet. It will cross within Neptune's orbit and reach perihelion in 1989. The unusual features of Pluto may be explained by the suggestion that it was at one time a moon of Neptune.

The Minor Planets

There is one more planet that can sometimes be seen with the unaided eye under excellent conditions. This is a very tiny one, the brightest asteroid, Vesta. The asteroids form a group of thousands of small fragments of rock circling the sun as planets do. Their orbits lie mainly in the region between Mars and

Jupiter. Some of the asteroids are so small as to be called "flying mountains." A few decades ago the derogatory term "Vermin of the Skies" was applied to them, because of the nuisance of keeping track of them. Now their image has brightened and they are being designated as the Rosetta Stone of the solar system, because they could help solve the mystery of its origin.

The brightness of Vesta at different oppositions varies considerably, as shown in the table by Jean Meeus, with the dates and magnitudes of all oppositions for the rest of this century. At some oppositions it is visible to the unaided eye. Not till 1982 will we have Vesta as bright as in 1974, with magnitude 5.8. Its distance from the sun is 2.36 times that of the earth. It takes 1,325 days to orbit the sun, and comes to opposition about three times in four years. Its path in the months around opposition is always given in *The Observer's Handbook*.

Oppositions of Vesta

Date	Mag.	Date	Mag.
1975 Sept. 18	6.1	1988 Jan. 22	6.4
1977 Jan. 9	6.5	1989 June 26	5.5
1978 June 5	5.5	1990 Nov. 15	6.6
1979 Nov. 3	6.5	1992 Mar. 9	6.0
1981 Feb. 21	6.2	1993 Aug. 28	6.0
1982 Aug. 10	5.8	1994 Dec. 25	6.6
1983 Dec. 13	6.6	1996 May 8	5.6
1985 Apr. 18	5.7	1997 Oct. 17	6.4
1986 Oct. 3	6.3	1999 Feb. 4	6.3

Jean Meeus in *The Journal of the British Astronomical Association*.

Although Vesta is much the brightest of the asteroids with a diameter of 300 miles, it is not the largest, which is Ceres, with a diameter of 620 miles. Pallas ties with Vesta for second place, with diameters of about 320 miles. Recent work indicates densities for Ceres and Vesta similar to carbonaceous chondrites, but Vesta has features in its spectrum not yet matched by any other asteroids. Like all bodies in the universe, the asteroids are

Plate 12. Left: star field with Nova Cygni absent. *Right:* Nova Cygni August 30, 1975, with 16-inch U. of T. telescope.

Plate 13. Crab Nebula in Taurus, in red light, 200-inch Hale telescope.

Plate 14. Schematic diagram of our galaxy, with globular star clusters (scale in parsecs).

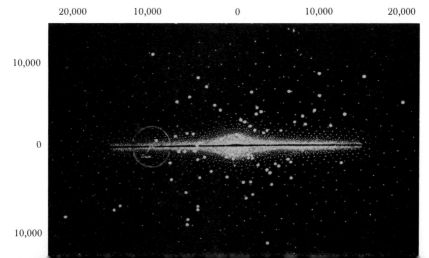

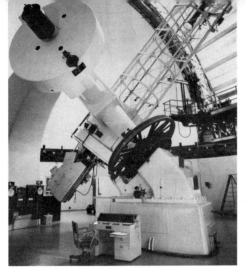

Plate 15. The 72-inch reflecting telescope, Dominion Astrophysical Observatory.

Plate 16. The 150-foot radio telescope, Algonquin Radio Observatory.

Plate 17. The U. of T. 24-inch telescope, Cerro Las Campanas, Chile.

Plate 18. Grinding the disk at the Dominion Astrophysical Observatory for the Canada-France-Hawaii telescope.

Plate 19. The Pleiades photographed by Canadian amateur astronomer Jack Newton.

Plate 20. Messier 13, the great globular star cluster in Hercules, taken by the author with 74-inch David Dunlap reflector.

Plate 21. Sagittarius clouds of the Milky Way, in red light, 48-inch Schmidt, Hale Observatories.

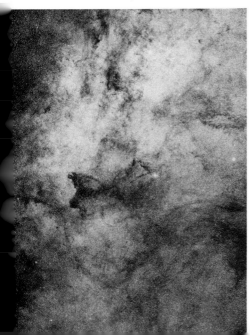

Plate 22. The southern globular cluster Omega Centauri, with U. of T. 24-inch, Chile.

Plate 23. The Large Magellanic Cloud, nearest galaxy to ours.

Plate 24. The Small Magellanic Cloud and globular cluster 47 Tucanae.

turning on their axes, Vesta in a period of 10.6 hours — longer than that of the giant planet Jupiter! Accurate measures of light variation indicate the rotation periods.

The discovery of the asteroids is another interesting chapter in the annals of astronomy. In mid-18th century J. D. Titius of Wittenberg was fascinated by the regular progression in the distances of the known planets from the sun. He developed a simple formula to represent them, in terms of the earth's distance from the sun.

Write a string of 4's. To the first add 0, to the second add 3, to the third 3 x 2, to the fourth 3 x 2^2, and so on. Divide by 10. Then what have you? The distances of the planets from the sun in terms of the earth's distance from the sun.

4	4	4	4	4	4	4	4
0	3	6	12	24	48	96	192
.4	.7	1.0	1.6	2.8	5.2	10.0	19.6
Mercury	Venus	Earth	Mars	Missing	Jupiter	Saturn	Uranus

The table gives the relations as known in the 18th century.

At 2.8, between Mars and Jupiter, no planet was known. After the law seemed to be confirmed by the discovery of Uranus in 1781, the prominent German astronomer Johannes Bode decided a search for the missing planet should be begun. It was undertaken by a group of 24 European astronomers, dubbed the "Lilienthal Detectives," because the plans were laid at the Lilienthal Observatory near Bremen.

Before the detectives had a chance to take much action, the missing planet was found independently by a Sicilian astronomer, G. Piazzi, making a systematic search of the sky at Palermo. On the first night of the 19th century, January 1, 1801, he discovered an object which, from its motion, he recognized as the missing planet, though it was much fainter than he expected. He named it Ceres after the divine mythological protectress of Sicily.

To everyone's surprise, other missing planets were found: four were known by 1807. Vesta was discovered by Olbers on March 29, 1807. Then toward the end of the century the introduction of celestial photography revealed asteroids by the hundreds. An asteroid shows up on a photographic plate by a trail as it moves among the stars. The number whose orbits are known today approaches 1,600. The actual number of these fragments in space (some of which undoubtedly reach us as meteorites) must run into the hundreds of thousands.

With Vesta our roster of planets visible to the unaided eye comes to an end. We do not yet know whether all the sizable planets in our solar system have been discovered. But it is obvious that no others will have naked-eye visibility.

The two most recently found planets do not satisfy Bode's law well. The true distance of Neptune is 30.1, not 38.8, and of Pluto 39.5, not 77.2. Pluto is closer to the Bode's law value for Neptune.

Now that man's exploits in space have met with such amazing success, some scientists are beginning to cast eager eyes at asteroids as landing spots, or even for capture. We now have the necessary technology for landing. A hunt is on for the best candidate. Some of the thousands of asteroids with uncharted, highly elliptical paths may approach the earth within a few hundred thousand miles. An International Astronomical Union bulletin asks all observers to pay particular attention to any long asteroid trail on a photographic plate. A long trail means a fast-moving asteroid and hence a near one.

On an asteroid "man will look big and feel great" because he weighs only a fraction of an ounce. The gravitation on an asteroid is probably 1/10,000 that on earth. He won't need a vehicle to get him about. Even with a clumsy space suit on, he can jump half a mile high and 10 minutes later drift safely back to the surface. With his half-mile jumps he'll go round his little world in a couple of hours. But he needs to choose an asteroid more than half a mile in diameter. Then its gravity will be sufficient that he won't have the misfortune inadvertently to jump right off it.

The capture of an asteroid is a real possibility. Space technology should now be equal to capturing one with a mass of less than a few hundred pounds and bringing it safely back to earth, for earthbound humans to look at in wonder.

★

★

SEVEN

Comets

When beggars die there are no comets seen.
The heavens themselves blaze forth the death of
princes.

Julius Caesar, II, 2

Rated highest on the list of astronomical events that attract most public interest are comets. Recorded appearances of comets go back hundreds of years before the Christian era. In itself, the word "comet" is a very old one. It comes from the Greek *cometes*, meaning hairy, and was used originally with the noun *aster*, meaning star. The Greeks attributed the origin of this name of "hairy star" to the Egyptians.

Until four centuries ago comets were believed to be transitory, and largely atmospheric effects; perhaps some sort of terrestrial emanation. From the earliest times they have been regarded as omens of doom, though occasionally their bad influences seem to have been interspersed with good ones.

The stories of the fear and fanaticism that have come down to us, associated with comets, seem almost incredible until you make allowance for the fact that comets were an unknown and unexplained phenomenon on a gigantic scale. They brought forth all the instinctive fear of the unknown, which shows in very early records. Pliny took public reaction to comets into account when he enumerated 12 varieties, including one called "monstriferus," that is, horror-producing. Pope's translation of Homer's *Iliad*, Book XIX, 11, gives the impression of comets that prevailed for several millennia:

> *Like the red star, that from his flaming hair*
> *Shakes down diseases, pestilence, and war.*

And in fact, centuries later Milton's description of a comet (probably he was describing that of 1618) in the Third Book of *Paradise Lost*, is remarkably similar: ". . .and from its horrid hair shakes pestilence and war."

Many of the events with which comets were linked are well known in history. Word of the event itself has been handed down to modern times, but the cometary association is much less well known nowadays. The juxtaposition of the comet and the event in time is not necessarily close. Sometimes there is a spread of several years, but that does not seem to matter when you are looking for an omen.

Before the Christian era, the Romans thought the comet of 44 B.C. was a celestial chariot carrying away the soul of Julius Caesar who had been assassinated just before it appeared. Many omens, including a comet, are said to have preceded the death of the emperor Vespasian. When Vespasian overheard his courtiers discussing the comet as a threat to his life, he remarked, "This hairy star does not concern me. It menaces rather the King of the Parthians, for he is hairy, and I am bald."

However, when the great comet of 1680 drew much attention on both sides of the Atlantic, the Massachusetts preacher

Increase Mather issued warnings about it. He noted that when the Emperor Vespasian joked with the Parthian about the hairy star, it was Vespasian who died within the year, not the Parthian! Mather was not alone in his worries, for a medal was distributed by monks to avert misfortune caused by this comet.

It is easy to see how horror stories have come to be a part of our literature when you read some of the descriptions of early comets. The historian Nicetus described the appearance of the comet of 1182 A.D. thus: "After the Latins had been driven from Constantinople, an omen was seen of the rages and crimes to which Andronicus was about to abandon himself. A comet appeared in the sky: like a coiled snake, it sometimes stretched out and sometimes bent back on itself. Sometimes, to the horror of the onlookers, it opened an enormous snout, as though, greedy for human blood, it was about to drink its fill."

After eons of abysmal ignorance about the how and why of comets, understanding of them began to appear, first as a tiny spark, and then after some decades as a real flame. It began with Tycho Brahe and his discovery on November 13, 1577, of a comet that he observed for two months. Tycho found the comet before sunset, while he was fishing. After dark the tail was 22° long, red and curved in the twilight. Though Tycho had no telescopes, his massive instruments of position were sufficient to prove that the comet was farther away from earth than is our moon. Hence it could not be just an "exhalation" in the earth's atmosphere.

Tycho was the last great astronomer to live before the invention of the telescope, and one of the outstanding observational astronomers of all time. It was from Tycho's extensive series of planet observations that Kepler was able to derive his laws of planetary motions. Tycho was a very colorful figure, with a nose made of silver and gold. He had lost his original one in a duel at the University of Rostock, and he constructed the metallic one in its place. The duel was not over a woman, as one might presume, but over an argument as to whether he or his opponent knew more mathematics.

As the 17th century progressed, one astronomer after another increased the understanding of cometary paths. Finally Dörfel of Upper Saxony, after studying the great comet of 1680, proved that it was traveling in a parabolic orbit with the sun at one focus. Newton then computed that comets were moving in orbits that are really elliptic, like those of planets, but of such high ellipticity that they approach parabolas. He also developed a method by which the orbit of a comet could be computed from only three observations.

The crowning triumph was by Halley, who applied Newton's principles and laboriously computed the elements of the orbits of 24 comets. Three were so similar that he concluded they were one and the same comet. The first had been observed by Appian in 1531, the second by Kepler in 1607, and the third by Halley himself in 1682. Halley's name is attached to his famous comet, not because he was the discoverer, but because he was the first person who dared predict that a comet would return. In his papers to the Royal Society he wrote:

"For, having collected all the Observations of Comets I could I have fram'd this following Table, the Result of a prodigious deal of Calculation: Which, tho' but small in Bulk, will be no unacceptable Present to Astronomers. For these numbers are cable of representing all that has been yet observ'd about the Motion of Comets, by the Help only of the annex'd General Table; in the making of which, I spar'd no Labour, that it might come forth perfect, as a Thing consecrated to Posterity, and to last as long as Astronomy itself. . . . Hence I think I may venture to foretel, that it will return again in the year 1758 . . . Wherefore if according to what we have already said it should return again about the year 1758, candid posterity will not refuse to acknowledge that this was first discovered by an *Englishman.*"

Halley was in middle life when he made his famous prediction, and he could not hope to live to see it come true. But come true it did. Another link was fastened in our chain of understanding of the solar system when Halley's comet was first seen again on Christmas, 1758, by Palitsch near Dresden. It was visible for five months, with a tail 47° long on May 5.

Nature of a Comet

Science today takes a more dispassionate view of comets. With telescopes and spectrographs and ultimately space probes, astronomers are bent on finding out just what a comet is. It has been described as an airy nothing. By terrestrial standards — that is, if you had to cart its material around in a wheelbarrow — it has a lot of mass. By astronomical standards it has little. At present only an upper limit can be set to the amount of material in a comet because none has ever caused the slightest observable deviation in the movement of another solar system body. Extensive studies have been made to try to find such a gravitational attraction, especially at times when comets have been playing in and out amongst the moons of Jupiter. The mass of a comet is estimated to be less than 1/10,000 that of the earth. The total mass of Halley's comet was estimated by H. N. Russell as 25 million tons, and this would be about 5 percent of the amount excavated to make the Panama Canal. The average mass per cubic mile of the comet would be about half an ounce.

A comet consists of three main parts, only one of which is found in all comets. That is the fuzzy envelope known as the coma, which varies in size with each comet as well as with its distance from the sun. The ability of a comet to form a coma is the property that distinguishes it from an asteroid, for example. The coma is usually some tens of thousands of miles in diameter. In some comets a second part, a sharp nucleus, only a few miles or less in diameter, may be seen toward the center of the coma. Most comets have a tail, and sometimes this can be very spectacular. The tails of many comets are the largest things in our solar system, millions of miles in length. One of the longest was that of the great comet of 1811, variously estimated as between 100 and 200 million miles.

Comet tails are formed when the pressure of light from the sun and the solar wind drive gases from the nucleus into the coma and then into the tail. The tail of a comet is thus directed away from the sun. Centuries ago Seneca said, "The tails of comets flee from the sun's rays." As the comet approaches the sun it drags its tail after it, but as the comet leaves the sun's region, its tail precedes. In other words, the tail then wags the dog.

Some of the bright comets with very long periods pass very close to the sun at perihelion with enormous velocities. The comet of 1843 passed only 500,000 miles from the sun, traveling at 300 miles a *second*. Sometimes the tail of a comet becomes curved. Or sometimes the comet grows several tails, one after another, and they are all visible at once. This was the case with de Cheseaux's comet of 1744, which had six tails.

The material driven into the tail of a comet is lost to that comet forever. This is the reason why the comets with periods comparable with human lifetimes are mostly rather faint and inconspicuous. A comet literally wears itself out in returns to the sun, losing somewhere between 1 percent and 0.1 percent of its mass on every approach to the sun. Estimates are that a comet can return between 100 and 1,000 times to the sun before its volatile material is all driven off. It is the comets that come in only once in many hundreds or thousands of years which are really long-lived.

There are nearly 100 comets known with periods of less than 200 years. These are comets that have come under the control of the outer planets. Perhaps at one time they came in toward the sun on orbits that were nearly parabolic. But then one of the outer planets cast a gravitational spell on them and condemned them to oscillate between the sun and the orbit of that planet. Jupiter, as the most massive planet, is the most active spellbinder.

In Whipple's "dirty iceberg" theory, the volatile substances are all expelled from the comet and after a long period of time only solid mineral particles or hunks remain. The spectroscope gives us an analysis of the material in a comet, showing common elements and molecules to be present. A comet's nucleus is composed of ices of methane, ammonia and water, in which are embedded small particles of elements such as iron, nickel, calcium, magnesium, silicon and sodium. It is possible that some of the meteorites crashing down on earth are worn-out comet nuclei and not hunks from the asteroid belt.

Our solar system has now been in existence 4,500 million years. The reason that there are still many comets in existence, when they constitute such a perishable phenomenon, is one of the unanswered problems in astronomy. One suggestion is that there is a great reservoir of millions of comets out beyond the outermost planets. From time to time some of these are hurled inward toward the sun by passing gravitational forces acting on them.

Discovery of Comets

Systematic comet hunting has been a fascinating game ever since the invention of the telescope, and probably before. About 10 comets each year can be found with telescopes. Most of these are not new comets but returns of old ones.

The record for discovery of comets is held by Jean Louis Pons of Marseilles with 45 to his credit, from 1801 to 1827. His phenomenal discovery of comets earned him the nickname "Comet's Magnet." Another French astronomer, Charles Messier, at a slightly earlier period, had many discoveries though his chief claim to fame lies in making a list of celestial objects that observers should *not* mistake for comets, as discussed in Chapter 9.

The banner year for comet discovery so far is 1967, with 14 comets reported and confirmed. In 1947, 1948 and 1960, 14 comets had been provisionally accepted, but later on in each of these years one could not be confirmed. Included in the list for 1967 were 10 recoveries all credited to the diligent Japanese astronomer K. Tomita, who works with the 36-inch reflector of the Dodaira Station of the Tokyo Observatory. On October 5 that year Tomita achieved a unique feat. He recovered four known comets in one night. (An observer who first locates an old comet on its next return is said to have recovered it.) Comets have several designations. Besides bearing the names of the discoverer, they are assigned numerals in order of their closest approach to the sun; i.e., 1967 III.

Beginning in the last century, American astronomers have many comet discoveries to their credit, some of which have received wide publicity. In 1831, the King of Denmark, His

Majesty Frederic VI, established a gold medal for the discovery
of a telescopic comet. On October 1, 1847, Miss Maria Mitchell of
Nantucket discovered such a comet. The discovery brought the
first award of this medal to an American, and its first award to a
woman anywhere in the world. The Maria Mitchell Observatory
on Nantucket Island, founded in her memory, continues to make
an astronomical contribution in both research and public nights,
when thousands of people have been brought closer to the stars.
This is an appropriate sequel to the Latin inscription on Maria
Mitchell's medal, translated as "Not in vain do we watch the
setting and rising of stars."

Although he is not credited with the largest number of
discoveries in America, the best-known American comet seeker to
date was E. E. Barnard. His life underlines what persistence and
devotion to astronomy can do. Born at Nashville, Tennessee, in
1857, he was fatherless by the end of the Civil War. All his
education was given him by his mother, except for two months at
school. When only eight or nine he went to work in a
photographic studio in Nashville. In 1876 he became so
fascinated with a popular book on astronomy that was loaned to
him that he mounted a one-inch lens in a tube. Then by
scrimping and saving he was able to purchase a five-inch
telescope. In 1881 he began to search for comets and eventually
over the years found a total of 19. They certainly came in handy,
because in 1880 Mr. H. H. Warner, founder of the observatory
bearing that name, offered a prize of 200 dollars for every new
comet found by an observer in Canada or the United States. As
one by one the comets obligingly appeared, Barnard used his
comet money to construct a house for his mother and his wife,
one he referred to as "the house that was built with comets." His
total of 19 new comets was slightly exceeded by Professor W.H.
Brooks of Geneva, New York, who found 20.

An almost unbelievable coincidence in comet discovery
occurred in 1896, and is well documented in scientific journals by
well-known astronomers. On February 14, 1896, the Lick
Observatory received a cable from an astronomer at Kiel,
Germany, stating that he had reobserved a comet discovered
months earlier by C. D. Perrine of Lick. When Perrine checked

the cabled position, he saw that it could not be his comet. The skies were clear the next morning and he turned a 12-inch telescope to the position. There was an 8th-magnitude comet. He assumed it had been discovered at Kiel. It took some weeks to detect that, in decoding the cable, an error had been made in the position. Turning the telescope to an erroneous position, Perrine had actually discovered a new comet! Furthermore, the comet was moving so rapidly that only on the very morning Perrine looked for it would it have been in the field of his telescope in that position! It was not there when the cable was written, nor on the day the position was handed to Perrine, nor would it have been there on the day following his discovery.

Halley's Comet

Few persons now can remember from first-hand experience any of the comets appearing earlier than the 20th century. But many still have vivid recollections of some of the bright naked-eye comets of this century, particularly Halley's. Halley's was not the only naked-eye comet visible in 1910 — there was another daylight comet that year — but Halley's is the one best remembered. And as the date of the next return of Halley's comet in 1985 approaches, interest in this and other comets can be expected to undergo a great flare-up.

The period of Halley's comet is around 75 years, but varies up to 77 years because of planetary perturbations. After Halley

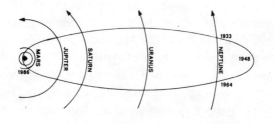

Figure 20. *Orbit of Halley's comet.*

showed that three comets which appeared about 75 years apart were one and the same comet, its appearances could be traced back. The fact that Halley was not the discoverer of this comet is often hard to get across. On examinations my students have sometimes written "Halley founded Halley's comet" in their gropings as to just what Halley did do. Professor Alice Farnsworth of Mount Holyoke College constructed a jingle to help her students remember the pertinent points of this comet.

> *"Of all the objects in the sky*
> *There's none like Comet Halley.*
> *We see it with the naked eye*
> *And periodically.*
> *The first to see it was not he*
> *But still we call it Halley.*
> *The notion that it would return*
> *Was his originally."*

Since 240 B.C. records of every return have been found. In 11 B.C. Halley's comet appeared high in the zenith at Bethlehem, in the land of Judaea. There is speculation as to whether or not Halley's comet could have been the Star of Bethlehem. The exact date, and even the year, of the birth of Christ are unknown, but 11 B.C. is somewhat earlier than the limits usually accepted for this important date. However, the Star might still have been a comet because a bright comet did appear in 4 B.C., a date that is probably closer to the true value of the beginning of the Christian era.

Just as for other comets, the appearances of Halley's are frequently associated with historical events. Amongst the recorded returns, in April 6, 218, "a fearful flaming star" preceded the death of Emperor Macrinus in Rome, and on July 3, 451, one preceded the death of Attila.

One of the most famous appearances of Halley's Comet is that of 1066, when it was recorded for posterity in the Bayeux tapestry. This tapestry derives its name from the fact that it was commissioned by Bishop Odo of Bayeux in Normandy and has for a long period been in the custody of the Bayeux civil authorities. It was worked by Queen Mathilda, the wife of William the Conqueror. It consists of 72 scenes on a strip of linen 20 inches wide and 231 feet long, worked in eight colors of worsted. There are two sections to the Halley's comet scene, *"Isti mirantur stellam."* One shows William the Conqueror and his courtiers rejoicing at the omen, and the other portrays Harold quaking on the throne.

In 1066 the comet was first observed April 2 by Chinese astronomers, and independently the next morning by Japanese, in Pegasus with a thin tail pointing southwest toward Aquarius. A European record of April 5 exists in the archives of the cathedral at Viterbo, Italy. On April 24, 1066, the comet passed to the evening side of the sun. This was the beginning of a week of clear weather on both sides of the English Channel, a somewhat unusual event. The comet was a magnificent sight for all, close to the western horizon. The Normans were convinced that its head and curving tail pointed downward to Harold's castle in Britain. Taking this as an omen for the cause of Duke William, they developed a crusading spirit with a rallying cry: *"Nova stella, novus rex!"* ("A new star, a new king.") William thought comets appeared only "when a kingdom wanted a king." On the Saxon side of the Channel, despair prevailed. Those who believed in the prognostications of the comet had their ideas confirmed on October 14 when King Harold, his brothers Gyrth and Leofwine and the flower of the Saxon army lay dead on the battlefield 10 miles northwest of Hastings. A curious coincidence came centuries later, on June 7, 1944, when the first town on the Norman coast to be liberated by William's descendants was Bayeux.

Another dramatic period came at the return of the comet in 1456. Three years before, Constantinople had fallen to the Turks. Pope Calixtus III had tried to organize a crusade, but he was weak and mentally incompetent and failed to receive support. Apparently the comet terrified him. He is said to have issued a papal bull against it. It is certain that he ordered the Ave Maria repeated at midday — the beginning of the noon Angelus. He also added a new clause to the regular prayer so that it read, "Lord save us from the devil, the Turk, and comet." The comet that year was a truly magnificent object, with a tail extending a third of the way across the heavens.

At its last return, Halley's comet was first found by photography by Max Wolf of Heidelberg on September 11, 1909, though subsequently it was shown that at Helwan, Egypt, a picture of it was taken August 24. It was still more than 300 million miles from earth when first discovered. On subsequent nights it was found at the Lick and Yerkes observatories. From then on it could be observed till July 1, 1911. From November, 1909, to February, 1910, it brightened by at least two magnitudes. On the early photographs the comet does not look fuzzy, but rather just like a faint star. The second week in May, 1910, it reached its greatest size and brilliance. On May 18 its tail was 105° long, but 30° of this was below the horizon. On that evening the head of the comet passed between the earth and sun. On the following day the earth passed through the tail. The only visible results of this were a persistent nocturnal glow that lasted through the night in some places, and a few more meteors than usual.

However, there was great apprehension among large sections of the populace when it was learned that the earth would pass through the tail of the comet. It is true that comet tails do contain some noxious molecules like cyanogen, but the amount of such material is much too small to have any ill effects on life.

It is not surprising that ignorant people in centuries past went into a great stir of panic at the appearance of a comet. It is surprising, however, that this has happened in our own century and on our own continent. Dr. Dinsmore Alter of the Griffith Planetarium, Los Angeles, made a collection of newspaper reports at the time of this last appearance of Halley's comet, and they make revealing reading indeed. Alter notes that such an event makes a disturbance natural for "the great lunatic fringe," and it is played upon by charlatans always eager to make an easy dollar.

From Haiti comes the following:

"FORTIFIED AGAINST COMET
Shrewd Voodoo Doctor in Hayti
Sells Pills to Ward Off Danger

"Whatever the comet may do to this old earth, the natives of Port Au Prince, Hayti are prepared. They are confident they will go unscathed, because they are well packed with comet pills.

"Officers of the Hamburg-American liner Allegheny said all the stevedores, servants, laborers, merchants, beggars, and thieves are rushing pell mell to the hut of a shrewd old Voodoo doctor just outside the city, who is selling comet pills as fast as he can make them."

You do not have to go as far as Haiti, however, to get into the comet-pill racket. Galveston, Texas, had it too.

"DO BIG TRADE IN COMET PILLS
Two Slick Crooks Swindle Texans

"Two men who sold hundreds of comet pills and mouth inhalers to superstitious Texans in four counties of South Texas were arrested in Barmoria county yesterday. The authorities had clear cases against the men for swindling, but the Courthouse was soon besieged by people begging for the release of the men.

"The pills and powders were a harmless preparation of sugar and quinine and the purchasers were told that the effect of the prescription would prepare their systems to withstand the gases of the comet's tail. The leather inhalers to fasten on the mouth were also to protect the wearers against the gases. The victims paid from $5 to $10 for the layout, some taking as many as fifty, and from $2 to $24 for the inhalers, which were to be worn when asleep.

"The two men, claiming to hail from Ohio, had many agents working and they collected several thousand dollars."

Also in the southern United States, conjure bags and work strikes made their appearance. In Georgia:

"BUY CONJUR BAGS

"Dealers in conjur bags carried on a thriving business today as the result of the scheduled trip of the earth through the tail of Halley's comet tonight. Meetings also were held in the churches today, thousands refusing to return to work until the passing of the comic [sic]."

In Texas the work strike was having serious consequences:

"PRAYERS AND LAMENTATIONS

"Thousands of Texans, fearful that the end of the world will be produced by the comet striking the earth, are quitting their labors and gathering in churches and camp meetings. Many are flocking from the country to the small towns and in congregations devoting their time to prayer and lamentations. . . .

"Throughout the lumber camps in East Texas, where thousands of laborers are employed, about ten timber-cutting camps have closed down for want of workers."

In Kentucky also they prepared to meet the end of the world.

"EXPECTING WORLD TO END — ALL NIGHT SERVICES

"In Lexington excited people are tonight holding all-night services, praying and singing to prepare themselves to receive the celestial visitor and meet their doom."

Worry about the comet appears to have driven some people to suicide and others out of their mind.

And of course, amidst all the scare, there is inevitably the practical joker. This one contrived a spectacular act, of which one now reads with amusement.

"CHEMIST PLAYS TRICK

Terrorizes Whole Town

"Comet watchers of Roselle, New Jersey, were thrown into a state of terror last night as the result of the practical joke of Arthur Smith, a chemist, of that place. Knowing that the residents of the town were on the lookout for any phenomena that might result when the earth passed through the tail of the comet, Smith, with the aid of a small balloon, a quantity of sodium, a time fuse, and a stick of dynamite, contrived an apparatus which would rise into the air to a height of 1000 feet and then explode with a terrific roar, igniting the sodium, which would fall to the earth in a great shower of flame. Smith, accompanied by his son, took the apparatus to a large vacant field in the south section of the town. It was released and worked to perfection. The explosion of the dynamite could be heard for miles. Fully an hour passed before fears of the people were allayed."

Halley's comet will next return in 1985-86. Toward the end of 1984 it should be seen again on photographs with big telescopes. It will probably first become visible in large telescopes in the morning sky of August, 1985. By mid-November it is expected to be magnitude 7 or 8 in Taurus, with the beginnings of a tail. In January, 1986, it will be in the evening sky in Aquarius and will disappear in the twilight around January 24 at about 4th magnitude. After perihelion about February 9, 1986, it will reappear in the morning sky, but it will then be better viewed from the southern hemisphere. It is expected to be at its best in March with a tail 20° long, even as much as 40°, but well visible only from the southern hemisphere. During the last few days of April, 1986, the comet will have moved northward into Crater. This may be the best view for northern hemisphere residents, and the comet may be of magnitude 4 or 5 with a tail 5° long.

Even a decade ahead, however, astronomers are sounding notes of warning that Halley's comet may next time prove to be a disappointing sight. The night sky illumination over the major cities of North America increased tenfold between 1961 and 1970. By 1985 if this keeps up, many large cities will have a permanent twilight. Only the brightest stars and planets will be visible from them. However, if you are away from city lights, and at a good altitude, especially in the southern hemisphere, Halley's comet may again provide the thrill that it has for myriads of people in several millennia.

Other Comets

Since the last appearance of Halley's comet, the northern hemisphere had been rather bereft of bright comets until the century was more than half over. For some years astronomers had been saying that, by comparison with previous centuries, a bright comet was long overdue. This dearth of comets was suddenly remedied in 1957 with the appearance of not one, but two naked-eye comets in one year — indeed within five months.

The first and best known of these was Comet Arend-Roland, which was discovered actually in 1956 on November 8 at Uccle Observatory in Belgium by Sylvain Arend and Georges Roland, as a 10th-magnitude object in Triangulum, heading for perihelion in April. It became a naked-eye object toward the end of February, but it was then too close to the sun to be seen. After its perihelion on April 7 it moved rapidly northward into the evening sky and became the first comet that many people had been able to see easily and enjoy in several decades. On April 23 Dr. G. van Biesbroeck with the 82-inch McDonald reflector estimated its tail as 25° long, pointing north into Cassiopeia. Its total magnitude was 3, that of its nucleus 6. It surprised everyone with a slender jet 4° long pointing *toward* the sun, when for years astronomers had been saying that a comet's tail always points away from the sun.

After Comet Arend-Roland behaved in this manner, other instances were dredged up from the literature of comets that also presented this unorthodox appearance. And later comets such as Seki-Lines 1962c also have shown an antitail. Because of the high northern declination of Comet Arend-Roland, it became almost circumpolar for some nights in mid-northern latitudes and was enjoyed by vast numbers of people.

Comet Arend-Roland was still a lively conversation piece when without warning in early August, 1957, Comet Mrkos appeared on our terrestrial scene. It was officially discovered on August 2 by the Czech astronomer Antonin Mrkos, of 3rd magnitude in Cancer. Later it was shown that Airline Pilot Peter Cherbak had discovered it at 4 A.M. on July 31 M.S.T. over North Platte, Nebraska, at 20,000 feet, the first comet to be discovered from an aircraft. And, in fact, even earlier, Japanese observer Sukehiro

Kuragano at Yokohama had found it July 29, but neither of these discoveries had been confirmed and reported through official channels. Like enemy planes during the war, it had approached us nearly in line of sight with the sun, thus causing its sudden appearance. By a curious coincidence it was in the same part of the evening sky as Comet Arend-Roland, although the orbits of the two comets are very different. Mrkos had a remarkable tail, markedly different from night to night. By August 13 it was $12\frac{1}{2}°$ long and a second tail appeared, with fanlike streaks. Sodium emission lines were an unusual feature in the spectrum of the comet.

After these comets had faded, things were relatively quiet on the bright comet front until 1965. Then on the morning of September 18 two Japanese comet seekers, K. Ikeya and T. Seki, found an 8th-magnitude, condensed but tailless comet just west of Alpha Hydrae. The brilliant and short career of Comet Ikeya-Seki received much publicity. It was in the morning sky when discovered. Observations showed it would cross into the evening sky for a few hours on October 21 as it swung around perihelion. Indeed in six hours it swung 250° around the sun as seen from earth. Though expected to be bright, the comet confounded the forecasters and proved to be nothing of an object that night. But the next day thousands of people, particularly in the higher altitudes of the southwestern United States, were able to see the comet in full daylight, near the sun, simply by holding up a hand to block out the sun.

The comet developed a coma thousands of miles in diameter and in the early-morning sky its twisted tail was 25° in length. This proved to be the fourth longest comet tail yet on record, 70 million miles, exceeded only by the comets of 1843, 1680 and 1811. The tail was not spectacular on the approach to the sun, but became so on the outward journey, resembling that of the comet of 1680 in this regard. On November 5 this comet followed the pattern of other sun-grazers and broke in two, probably the result of unequal heating when near the sun. One fragment was starlike, the other fuzzy. It is a member of the sun-grazing family that includes the comets 643 I, 668, 1843 I, 1880 I, 1882 II and 1887 I.

A brilliant comet, Comet Bennett 1970 II, was found by John

C. Bennett, an active amateur astronomer of Pretoria, South Africa. After 333 hours of comet sweeping, he found it 15 minutes after he started on December 28, 1969. It reached perihelion on March 20, 1970. This was the best comet for amateurs to study in many decades — in a dark predawn sky at a good altitude. Called "The Great Comet of 1970," it showed two types of tails: Type II a bright, broad, curved dust tail 19° long to the naked eye on April 8, with nitrate dust identified, and a Type I faint, narrow straight tail 8° long on the same date.

The comet with the greatest fanfare and ballyhoo to date is Comet Kohoutek, 1973f. It was discovered on March 7, 1973, by an unassuming Czech astronomer, Lubos Kohoutek, working at the Hamburg Observatory. Less than two weeks before, he had discovered Comet 1973e, which was a more nondescript comet. Early orbit computations for 1973f showed that it would not reach perihelion till late December. Its discovery so many months before led to the belief that a really spectacular comet was on the way and newspapers, periodicals, radio and TV went all out in their publicity of "The Comet of the Century," though cautious voices were raised all along the way. Cruises and special expeditions to preferred localities were arranged. As thousands of people tried in vain to see the comet with their unaided eyes, either in the dawn sky before the perihelion of December 28, or in the early-evening sky thereafter, many dubbed it "The Flop of the Century" or at best, "The Shy Comet."

But for scientific observations it was not disappointing. Never before has such an array of sophisticated equipment been set up to observe a comet, from the astronauts on Skylab to orbiting satellites and the large ground-based telescopes. And vital information has been obtained, from the presence of curious compounds like methyl cyanide to the first proof of water vapor in a comet, thereby lending support to the "dirty iceberg" theory. At any rate, those who missed Comet Kohoutek on this round will never have another chance. There are various estimates of its period, of which the shortest is 70,000 years.

A faint comet discovered in 1974 has proved to be a record breaker in two respects. It is the first new comet to be found by a Canadian astronomer, and it is the most distant comet yet found. On the night of November 12, 1974, Dr. Sidney van den Bergh of the University of Toronto was photographing the spiral galaxy in Triangulum with the 48-inch Schmidt telescope of the Hale Observatories on Palomar Mountain. When he developed the plate the next day, there was a faint comet of the 17th magnitude! After weeks of observations, the orbit as determined by Dr. Brian G. Marsden of the Smithsonian Astrophysical Observatory showed Comet van den Bergh (Comet 1974g) to have the greatest perihelion distance of any known comet. At perihelion on August 8, 1974, it was 560 million miles from the sun. The previous record was held by Comet Schwassmann-Wachmann (1) at 510 million miles. Comet van den Bergh will not return in the foreseeable future.

One inconspicuous comet, sometimes visible to the naked eye, deserves special mention. It is Encke's, whose period is the shortest known of any comet whose return can be predicted. (Comet Wilson-Harrington in 1949 with a shorter period of 2.3 years was observed for only six days, and is hopelessly lost.) The period of Encke's comet is 3.3 years, or 1,205 days. Pons at Marseilles, November 26, 1818, discovered it, but like Halley's comet it was named for the man who predicted its return. J. F. Encke, then only 27, identified its orbit with that of the comets of 1786, 1795 and 1805. By computation he proved that these were all one and the same comet. (The comet had also returned to perihelion seven times between 1786 and 1818 without being noticed.) He predicted its return in 1822, and named it Pons' comet, but everyone else called it Encke's comet, a name that it still bears.

In 1820 Encke suspected that the comet was gradually accelerating and that its period was diminishing at the rate of 0.11 days per revolution. When Encke died in 1865, F. E. von Asten, an astronomer at Pulkova Observatory near Leningrad, took over the safekeeping of the slightly erratic Encke's comet. Unfortunately the comet began to be less than cooperative, showing a sudden decrease in its acceleration in 1868 and a complete disappearance of it in 1871. It is said that worry over the comet's behavior was a cause of Von Asten's early death at the age of 36.

Encke's comet reached naked-eye brightness, magnitude 4.6, on its 1964 return, on June 4. At its last perihelion passage on

April 28-29, 1974, it did not reach naked-eye brightness. Before its last return it was photographed as a 20th-magnitude object through its aphelion, by Dr. Elizabeth Roemer of the University of Arizona. It is a rare feat to follow a comet through aphelion.

Comet Probes

As our space probes extend to actual contact with more and more celestial bodies, comets become an obvious target for one. Probably a comet probe is the best, perhaps the only, way to solve the mystery of comets. What peculiar property causes them to emit a gaseous envelope, the coma, while asteroids in the same region of space do not? Whence have these hunks of frozen vapors come — from the outer planets or their satellites or far beyond them in the realm toward the stars? Theories suggest that the solar system debris, which forms into comets, may extend one-fourth of the way to the nearest star. With our present technology a comet probe cannot be long away, even though it is a more hazardous undertaking than a close approach to the moon or Mars. Encke's comet is a likely target for a probe.

It has even been proposed that an explosive charge be sent to the nucleus to stir it up and hence make a study of cometary effects quick and easy. Predictions are that Encke's comet is apt to die a natural death in a few decades anyway, from getting closer and closer to the sun, so some destruction of it in the cause of science is not so revolting as might first appear.

A more pleasing idea is that of putting a comet to some use. The idea of catching up to one and using it as a natural machine to do man's bidding is not new to the present century. Jules Verne in one of his less well-known novels, *Hector Servadoc,* about 80 years ago told of the adventures of a French soldier "who got scooped up by a comet." Mark Twain also described a ride on a comet in "Captain Stormfield's Visit to Heaven." This may be another of the ideas of science fiction writers that eventually come true!

★

EIGHT

Bright Stars
and Constellations

Sparkling in splendor the Kite and the Dipper
Crossed the black welkin, and Scorpio's star
Lit on the runway stag, herdsman and skipper
When I was dust, perhaps, bed-rock or spar.

EDITH WYATT, "The August Sky"

What a glorious feast for the eyes is the night sky with its sparkling lights. "If the stars would appear one night in a thousand years, how would men believe and adore!" wrote Ralph Waldo Emerson. And our appreciation is enhanced if we recognize some of the more important stars and configurations as celestial friends. There are many excellent books and atlases that are concerned almost solely with the recognition of the brightest stars and constellations. But some discussion of bright stars and constellations naturally falls within the scope of this book.

To become familiar with many of these is not so difficult a task as many people think. It is a sad fact that many persons have been discouraged from further looks at the sky by futile attempts to see the old mythological figures in the heavens.

The stars visible to the unaided eye are nowhere near as countless as many people imagine. Years back I heard a famous movie magnate on a radio program declare solemnly that the users of his sponsor's product were "as numerous as the stars one sees when one steps outdoors on a clear California night." The number of 2,000 is considered a reasonable limit for the stars one could see with the unaided eye above the horizon at one time, under good conditions. A total of 6,000 stars visible to the unaided eye for the whole sky is the usual estimate. The product sponsored by the gentleman of the radio would long since have slipped into desuetude (which it has not) if its users were so few in number.

From the latitude of Toronto or New York there is a total of about 4,500 stars in all seasons you can see with the naked eye. Actually this number does not change greatly (except with altitude) for any place in the middle latitudes. Nearer the equator it increases because the entire sky becomes visible. Toward the poles in both the Arctic and the Antarctic it decreases, because at the poles one-half the sky is forever invisible. From the heart of big cities with bright lights around, probably only about 250 are easily visible. This number is gradually diminishing as the sky brightness over North America increases.

Bright Stars

Of these thousands of stars there are perhaps several dozen that are readily recognized by amateur students of the heavens. These are the stars bright enough to be noticeable, or those that are set in conspicuous configurations which, once you have learned them, you see again and again like a familiar friend. The bright stars have names derived mainly from Greek and Arabic. They also have another, frequently used designation from a 17th-century atlas of Bayer. He assigned to the naked-eye stars letters of the Greek alphabet in approximate order of brightness. When these are used, they are followed by the genitive of the constellation name. Alpha Orionis, then, is the brightest star in Orion; it is also known as Betelgeuse, meaning "the Armpit of the Central One." The 20 brightest stars in the sky are listed in the table below.

The Twenty Brightest Stars Beyond the Sun
(in order of decreasing visual brightness)

	Name	Constellation	Visual Magnitude
	Sirius	Alpha Canis Majoris	−1.4
n	Canopus	Alpha Carinae	−0.7
n	Alpha Centauri,d	Alpha Centauri	−0.3
	Arcturus	Alpha Bootis	−0.1
	Vega	Alpha Lyrae	0.0
	Capella	Alpha Aurigae	+0.1
	Rigel	Beta Orionis	+0.2
	Procyon	Alpha Canis Minoris	+0.4
n	Achernar	Alpha Eridani	+0.5
n	Hadar,d	Beta Centauri	+0.7
	Betelgeuse,v	Alpha Orionis	+0.7
	Altair	Alpha Aquilae	+0.8
	Aldebaran,v	Alpha Tauri	+0.8
n	Acrux,d	Alpha Crucis	+0.9
	Antares,v,d	Alpha Scorpii	+1.0
	Spica	Alpha Virginis	+1.0
	Fomalhaut	Alpha Piscis Austrini	+1.2
	Pollux	Beta Geminorum	+1.2
	Deneb	Alpha Cygni	+1.3
n	Beta Crucis	Beta Crucis	+1.3

n: not visible from Canada.
d: double star, combined magnitudes.
v: variable star.

You can gain a familiarity with the stars by identifying only a dozen or so of the brightest, which will serve as guideposts around the sky. It is like learning a big city, when you try to learn a few key points by means of which you can maneuver around. With the help of a star map you can learn others if you need to. It is not necessary to know every street.

Constellations

The idea of constellations goes back at least 2,000 years. Originally, a constellation meant a grouping of stars. Any stars in the area of the constellation that did not fit into the pattern were designated as "unformed." In early times, with imagination and superstition, people saw the shapes of animals or mythological figures in these groupings in the sky. In fact, the animals represented in the constellations give a clue as to the region where the system of naming them originated. It was not in India, because there is no constellation named for a tiger or elephant, nor in Egypt because there is no crocodile in the sky. Rather, the naming is thought to have started in the Mesopotamian valley.

Today, however, the word "constellation" has a different significance. It is a definite region of the sky, and every object within that region belongs to that constellation. Constellations might be considered comparable with states of the United States or provinces of Canada. There is a difference though, in that the entire surface of the heaven, all 41,253 square degrees of it, is divided up into constellations whose boundaries are established by the International Astronomical Union. There are now 88, the increase in number from earliest times being due largely to those added after exploration of the southern hemisphere first began. There are 18 constellations not visible from the latitude of New York City.

Magnitudes

The earliest star catalogues of Hipparchus (150 B.C.) and Ptolemy of Alexandria (150 A.D.) grouped stars in constellations. They also assigned brightness intervals, known as magnitudes, for the individual stars. After photometers (light-measuring devices) were invented a century ago, it became apparent that the division into magnitudes had a real meaning in terms of brightness. The brightness difference between magnitudes is approximately 2.5, and it is a geometrical progression. That is, a star of the first magnitude is 2.5 times as bright as one of the second, and it is 2.5 x 2.5 (about 6) times as bright as one of the third. The numbers 2.5, 6, 16, 40 and 100 are simple to remember for interpreting the difference in brightness between stars which differ by 1,2,3,4 and 5 magnitudes respectively. (For larger differences you simply continue on, 2.5 raised to the power of the difference in magnitude.) There are some objects in the sky brighter than the first magnitude, so the scale was extended backward into negative magnitudes. There are only 22 stars of the first magnitude or brighter, and these of course are the ones that mainly attract the attention of a casual observer. The current range in observable magnitudes is from the brightest object, the sun, at -26, to the faintest objects being studied with the largest telescope in the world, the 200-inch on Mount Palomar, at $+24$. This range is 50, equivalent to a brightness range of 2.5^{50} or 100^{10}.

Star maps fall into two broad categories. The equatorial and polar maps cover the entire sky. They have reference lines for the important coordinates of right ascension and declination.

Figures 23-26 show the constellations near the meridian for the four seasons for the observer facing south. The line (hour circle) above the date of observation corresponds to the meridian at about 9 P.M. Standard Time on the date indicated.

Figure 27 shows the heavens for an observer facing north in midnorthern latitudes. Again the vertical line through the center of the map — the north celestial pole — represents the meridian at 9 P.M. Standard Time on the date of observation. Turn the map counterclockwise for a later hour, clockwise for an earlier.

Figure 28 shows the region around the south celestial pole, invisible from Canada.

The second type of star map, called the horizon map, is constructed for an observer in (or near) a specific latitude. Loose copies of such maps will make a handy supplement to this book. Whatever map is used, it must be held so that the directions match the compass directions of the observer, and must be tilted to match the sky according to the date and time of night.

North Celestial Pole

In the northern hemisphere the place to start learning the stars is the northern sky. The projection of the earth's axis cuts the celestial sphere in the north at an altitude equal to the latitude of the observer. The altitude is the number of degrees above the horizon. For the northern U.S. and southern Canada this point is about halfway from the horizon to the zenith, the overhead point. It is called the celestial pole, and the whole sky seems to turn around it. Any stars that are no farther in degrees from the pole than the latitude of the observer in degrees will never set. These are called the circumpolar stars and their constellations are circumpolar. These stars are above our horizon all night every night in the year.

Among the circumpolar stars is the Big Dipper, certainly the best-known constellation in northern latitudes. Two factors make it well known. First, it is above the horizon every night of the year. Second, anyone can see that it looks like a familiar object, a dipper. It is represented on the state flag of Alaska. This constellation is also known as Ursa Major, the Great Bear. One of the mysteries relating to the constellations is why early peoples living in widely different sections of the earth — such as the Chaldeans and the Iroquois Indians of North America — saw in this group of stars the figure of a bear, a figure that few persons nowadays are able to imagine, even with conscientious trying. In autumn evenings the Big Dipper is to the left (west) of the Pole, and the Indian legend has the Bear going into his den for winter, heading down low under the Pole. In the spring, he re-emerges triumphantly, mounting up to the right and east. In England this constellation is also called the Plough or Charles' Wain, as Tennyson wrote in *New Year's Eve*,

> *"We danced about the May-pole and in the hazel copse,*
> *Till Charles's Wain came out above the tall white chimney tops."*

The Big Dipper is not merely an easy constellation to recognize. It is invaluable as a starting place to learn the sky. Its lead stars (leading in the direction of diurnal motion), known as

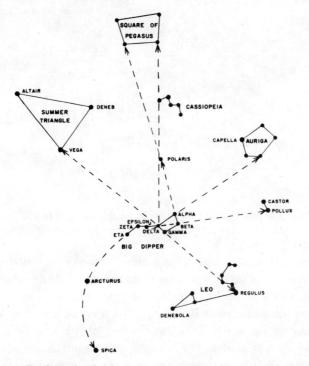

Figure 21. *Finding the sky from the Big Dipper. (By permission of Griffith Observatory.)*

"The Pointers," point to Polaris, the North Star. In addition there are certain key direction indicators in the Dipper for other portions of the heavens southward. The Pointers do not have an exact line of pointing. Precisionists will notice that the line does not reach Polaris *exactly,* but it is certainly near enough for purposes of identification. The Big Dipper is also useful in giving you a degree scale in the sky. The Pointers are 5° apart, and the top of the bowl is 10° wide.

When you follow along this line of sight and reach Polaris, you have moved into the constellation of Ursa Minor, the Little Dipper or Lesser Bear, in which Polaris is the brightest star, Alpha Ursae Minoris. Polaris itself, though an interesting star in its own right as a triple system, is only a 2nd-magnitude star and not unusually bright. Its importance lies, not in its brilliance, but in its unique position. It is very close to the north celestial

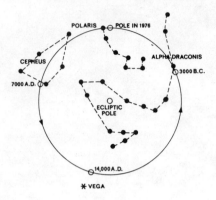

Figure 22. *The precessional path of the north celestial pole in a circuit lasting*
25,780 years. Different stars become Pole Stars.

pole around which our whole northern sky seems to turn.
Therefore it maintains a nearly constant altitude above the
horizon, day and night, year in and year out.

Polaris has not always been in such a favored position. We are
fortunate to be living at a time when there is a bright star
marking approximately the position of the north celestial pole.
(There is no bright star marking the south celestial pole.) Because
of the precession of the equinoxes, the north celestial pole
describes a circle with 23½° radius around the pole of the ecliptic
which is fixed, in Draco. It goes once around in 25,800 years. In
different millennia different stars earn the title of Pole Star.
When the pyramids of Egypt were building, about 3000 B.C.,
Thuban, Alpha Draconis, was the Pole Star. In early Greek and
Roman times there was no bright star marking the position of the
north celestial pole. Navigators had no such handy direction aid,
and had to use an entire constellation instead of one star. Some
used Ursa Major and some Ursa Minor. Even in the time of
Columbus, Polaris was not near enough to the Pole to mark its
position readily.

Alpha Ursae Minoris, now about 1° from the pole, has held the
post of Pole Star for only a few hundred years. The Pole will
move closer to Polaris until about 2100 A.D., when it will be half
a degree from it. Many centuries after that there will be no good
north pole star. A spectacular time to live on earth (in the

northern hemisphere) will be 14,000 A.D., when brilliant Vega will be the Pole Star, though at some distance from it. The Pole Star as a symbol of constancy can hardly be considered reliable.

Polaris is the star at the end of the handle of the Little Dipper. On a hazy night you may be able to see only two more of the stars in the constellation, Beta and Gamma. These form one side of the dipper, and are called "the Guardians of the Pole" because they are the nearest bright stars circling it. Alpha and Beta are 2nd magnitude, Gamma is 3rd, and no other stars in the constellation are brighter than 4th. So on many nights of the year you can see only these three stars in this constellation.

A third circumpolar constellation more conspicuous than Ursa Minor, is Cassiopeia, frequently called Cassiopeia's Chair because it resembles a chair in shape. It also looks like the letter W in the sky. Ursa Major and Cassiopeia are on roughly opposite sides of the Pole. When Ursa Major is high above the pole, Cassiopeia is low in the sky, and vice versa. In 1572 one of the most famous stars in the sky flared up in Cassiopeia. This was Tycho's nova, described in Chapter 9.

From mid- to high northern latitudes these circumpolar constellations are visible every night at varying altitudes above the horizon. It is interesting to study their positions some good evening several hours apart, to see how diurnal motion keeps them swinging counterclockwise around the Pole. A camera set up and pointed at Polaris (from a dark location) and allowed to run an hour or more will photograph star trails.

Most of the stars are more seasonal in their appearance in the sky. During hours of darkness they may be above the horizon for many months of the year, or for only a few nights. The amount depends on two things: their location with respect to the celestial equator and the latitude of the observer. Those that just miss being circumpolar will be visible almost every night in the year. Those on the equator will be visible for six months. Those which at best barely rise above the horizon will be visible only a few nights of the year for a few minutes.

Spring Sky

The first balmy spring evenings in April often tempt people to start learning a few stars. Sinking into the west then is the most magnificent constellation visible from mid-northern latitudes, Orion the Mighty Hunter. It is the only constellation visible from Southern Canada with two 1st-magnitude stars, red Betelgeuse and blue Rigel (Leg of the Jauzah). During the winter months Orion has been the dominant constellation, attracting attention with the symmetry of its blue-white stars, especially the even, linear spacing of the three that form the belt, spanning 3°. In the sword hanging down from the belt is one of the most beautiful objects in the heavens, the Great Nebula in Orion, described in Chapter 11.

Taurus, the Bull, is also then in the west, and north of Orion. As it heads toward setting, its characteristic V is upright. During the dark nights of World War II this V seemed to be a heavenly representation of Churchill's famous "V for Victory." The orange star in the V is Aldebaran, the eye of the Bull. In the V is the Hyades, a famous star cluster. The constellation of Taurus also contains the beauteous little group, which is even more famous, the Pleiades. Both of these clusters are discussed in Chapter 12.

Farther east from Orion, and higher, is a large rectangle. This is Gemini, the Heavenly Twins, with the bright stars Castor and Pollux. Pollux is yellowish, and slightly the brighter. Castor, with a bluish tint, is actually a system of two stars going around a common center of gravity, and the first such system for which orbital motion was proved. Any time we use the expression "By Jiminy" we are making a certain obeisance to this constellation!

South and east of Orion is blue-white Sirius, the brightest star in the entire sky. Sirius is in Canis Major, the Greater Dog, the constellation associated with the term "dog days." In ancient times Sirius was overhead at the hot season of the year. Presumably the addition of its hot rays to those of the sun was considered responsible for this type of weather. Sirius was worshiped by the Egyptians because its rising in the east signaled the time of rising of the Nile, and thus warned them of impending floods. Sirius has always been called the Dog Star, perhaps because of the warning it gave.

For centuries navigators have used a group of bright stars known as the "northern six." When astronauts Wally Schirra and Thomas Stafford wanted to design a patch for the flight Gemini VI they picked this group with the connotation of six. Capella is at the top, above the heavenly twins Castor and Pollux, then Procyon, Sirius and the constellation of Orion. The astronauts had an amazing coincidence. When they first saw their rendezvous vehicle, Gemini VII, it was between Sirius and the twins, exactly where they had placed it on the patch.

Spring, too, means Leo the Lion rising in the east, with the two bright stars Regulus and Denebola. The western end of Leo is shaped like a sickle. Regulus, the brighter, is at the end of the handle of the sickle, whereas Denebola brings up the rear of the animal. Regulus is one of the four "Royal Stars," which were selected for their importance by the Persians 3,000 years ago. They were presumed to mark the four quarters of the heavens. At the time they were assigned, Regulus was near the summer solstice. Fomalhaut in Piscis Austrinus, Aldebaran in Taurus and Antares in Scorpius are the other three. To the east and south of Leo is Virgo, a constellation without a striking pattern. It has the 1st-magnitude blue star Spica, and is important as the constellation which contains the autumnal equinox, one intersection of the equator and ecliptic. Spica forms an almost equal-sided triangle with Denebola and Arcturus, which we will soon meet.

Summer Sky

By late spring, as summer nears, the constellations in the west, like Orion and Taurus, have drifted toward the sun, and new ones are appearing in the east. Boötes, the Charioteer, contains the 1st-magnitude orange star Arcturus. This is one of the easiest bright stars in the sky to identify, once you have learned the Big Dipper. If you follow the curve in the handle, away from the dipper, you come to Arcturus, no matter what position the dipper has with respect to the Pole. You can continue on with the curve. A neat slogan is, "Follow the arc to Arcturus, then spiral on to Spica." Of course when the curve indicates that Arcturus should be below the horizon, it is!

Close on the heels of Boötes is Hercules. Each is a large, rather sprawling constellation with two straight-sided geometrical figures. Hercules contains one of the finest star clusters we know, described in Chapter 12. Snuggling between Boötes and Hercules is a regular, rather small semicircle of seven stars, Corona Borealis, the Northern Crown. Most of the stars are of the 4th-magnitude, but Alpha, the brightest, is of the 2nd magnitude. It is known also as Gemma, and called the "Pearl of the Crown." In 1946 observers were startled to see a new star of the 3rd magnitude join this group for a brief while. It proved to be the type of star known as a recurrent nova. Years before, it had been tagged as variable in light, and called T Cor Bor. Not many of these recurrent novae are known. Someday it may flare up again and give patient watchers another thrill.

Still farther east from Hercules, and circling even higher in the sky, is Cygnus, the Northern Cross or Swan. The figure really does look like a large cross in the sky. Actually, you may see the long piece of the cross in two different directions, depending on your selection of stars. The usual way is to have Deneb, Alpha Cygni, the brightest star in the constellation, at the top of the cross, and Albireo, Beta Cygni, at the bottom. The cross, which is on its side when rising in the northeast, lies almost entirely in the rich clouds of the Milky Way, which are discussed in Chapter 11. Deneb is one of the most luminous stars known in the entire sky. We are seeing it at a great distance, and its unusually large luminosity makes an accurate distance determination difficult, but it is at least 5,000 light years away. The configuration of this constellation changed markedly overnight on the Labor Day weekend, 1975, with the unheralded appearance of Nova Cygni, 1975 (see Chapter 9).

Between Hercules and Cygnus, at about the same distance from the equator, lies a perfect gem of a constellation, Lyra the Lyre with the bright blue-white star Vega, the fourth brightest star in the entire sky. Lyra, known to the early Britons as King Arthur's Harp, consists of a neat little equilateral triangle with a rhombus hanging from it. One star, Zeta, is common to the two configurations. Vega is at another vertex of the triangle, and the third vertex is occupied by a particularly splendid double star, Epsilon Lyrae. A telescope is necessary to resolve this star and then it proves to be a double double. Beta Lyrae, on the same long side of the rhombus as Zeta, is a fascinating variable star whose light variations can be followed with the unaided eye and are described in Chapter 9.

Straight down to the equator from Lyra is the large constellation of Aquila the Eagle with the bright star Altair, attended on either side by two 3rd-magnitude stars, forming almost a straight line, very distinctive of this constellation. The three stars Deneb, Vega and Altair form a large and conspicuous triangle of 1st-magnitude stars, which is known as the "Summer Triangle."

East of Altair and its lined-up companions, and somewhat above them, lies perhaps the neatest constellation in the entire sky, Delphinus, the Dolphin, otherwise known as Job's Coffin. It is another rhombus of four stars of 4th magnitude, similar in brightness, with a fifth star to the south forming the end of the dolphin's tail. It is certainly not one of the most important constellations, and largely because of its small size, it contains few interesting objects. Nevertheless its shape is so distinctive that once you have learned it, you can never fail to recognize it.

South of the equator all summer long we can see the richest part of the heavens, in Scorpius and Sagittarius. Scorpius looks so much like a kite with its tail dipping down toward the horizon that, if you can see the constellation at all above the horizon haze and earthly lights, it makes a good guidepost in the heavens. Its bright red star is Antares. The Chinese called it the "Fire Star." Antares is a supergiant, one of the largest stars known, with a diameter of more than 300 million miles. In the position of our sun it would fill the entire orbit of Mars.

Sagittarius, the Archer, gives us another dipper in the sky. The constellation consists principally of two rectangles joined together by a central star. The eastern or left-hand rectangle has been known for centuries by the appropriate name of "The Milk Dipper," perhaps because it seems to be scooping the Milky

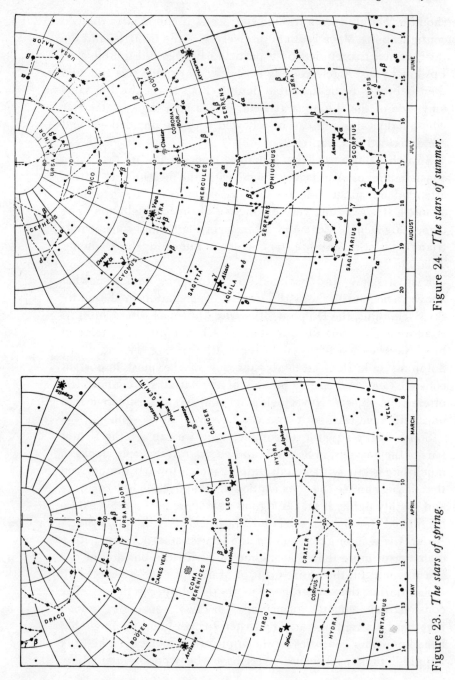

Figure 24. *The stars of summer.*

Figure 23. *The stars of spring.*

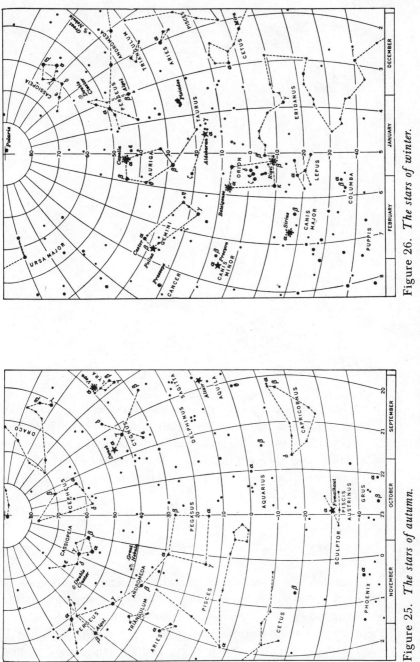

Figure 26. *The stars of winter.*

Figure 25. *The stars of autumn.*

(*Figures 23-28 are reprinted, by permission, from Baker and Fredrick, Astronomy, New York: D. Van Nostrand Co. 9th. ed, 1971*)

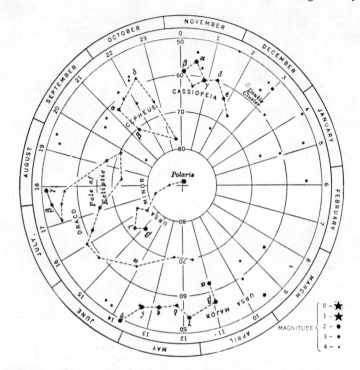

Figure 27. *The northern circumpolar constellations.*

Way. For in Sagittarius lie the heaviest star clouds of the Milky Way. From mid-northern or higher latitudes they are dimmed by their proximity to the horizon, but in the southern hemisphere, where they pass through the zenith, they are brilliant enough to cast a shadow. The course and significance of the Milky Way are described in Chapter 11.

Autumn Sky

Now as late summer comes on and autumn approaches, a different set of constellations begins to rise in the east in the evening. Slowly over the landscape rises the Great Square of Pegasus. You cannot mistake any other constellation for it, once all of the square is above the horizon. It is a real landmark among the stars, and knowledge of its position once saved the lives of the arctic explorer Donald McMillan and his party when the circumpolar constellations could not be seen because of cloud cover.

Pegasus, the Winged Horse, is also important as a guide to the intriguing constellation of Andromeda, in which lies the nearest of the great spiral galaxies to our own. Directions for finding it are in the final chapter. The easternmost or left corner of the square of Pegasus is actually in the constellation of Andromeda and is the brightest star, Alpha, in that constellation. Working from it you can follow the string of stars that make up the constellation. You can use Ursa Major as a guide to Pegasus and Andromeda also. A line from Delta Ursae Majoris, through Polaris and Beta Cassiopeiae, arrives at Alpha Andromedae, while a line from the Pointers of the Big Dipper through Polaris reaches the western edge of the square of Pegasus.

Winding under Pegasus is a rather inconspicuous, long and important constellation — Pisces the Fishes. Its chief importance lies in the fact that in this constellation the sun now crosses the equator going from south to north, hence it contains the vernal equinox. When astronomy began, two millennia ago, this crossing was in the zodiacal constellation next east, Aries the Ram, but precession of the equinoxes has caused it to slide westward into Pisces. The crossing is a little south and east of the septahedron known as the Western Fish.

As Pegasus, Andromeda and Pisces gain altitude in the east, you will see a bright curved string of stars in the northeast, the constellation of Perseus. This bears some resemblance to a reclining figure, perhaps stretching out to use the Pleiades as a footstool. Perseus is a constellation full of interesting objects, including the Double Cluster of stars, (Chapter 12) and the fascinating Winking Demon, Algol (Chapter 9). Perseus is one of the regions of the sky where bright blue stars congregate, and this shows up in the sparkling appearance of this constellation.

Close on the heels of Perseus, rising in the northeastern autumn sky come Auriga, the Charioteer, and Taurus already described. Auriga, like Perseus, is still in the Milky Way. It has a bright, distinctive figure, a pentagon of stars with the 1st-magnitude star Capella, the fifth brightest star in the entire sky. Capella is the same color as our sun, but the two are not similar because Capella is a giant with a diameter of about 14 million miles, compared to 865,000 miles for our sun. Capella means "She Goat," and the tidy little isosceles triangle hanging from the stars is known as "the Kids."

South of Auriga in the eastern sky is Taurus, with the V rising on its side, and the Pleiades above it. As Taurus rises, Aldebaran lives up to its epithet, "The follower of the Pleiades." Still farther south, somewhat low over the southern horizon, is Cetus the Whale. Just as the whale is the largest mammal, so Cetus is the largest constellation. For all its size, Cetus is not very conspicuous in the northern hemisphere, partly because it does not gain a high altitude, but mainly because it has no star brighter than 2nd magnitude. It consists of a pentagon to the east joined to a rectangle in the west by two stars, one of which is not always there. Delta is always present, but the other, Omicron, is Mira the Wonderful Star, which comes and goes, as discussed in Chapter 9.

A star that could become very significant for us is Tau Ceti, in the lower left-hand corner of the rectangle. This is one of the nearest stars, only 12 light years from us, and one that could have planets circling it. Tau Ceti was on the list for observation in 1960 when Dr. Frank Drake tried listening with the 85-foot dish of the National Radio Astronomy Observatory at Green Bank, W.Va., to see if he could detect signals that might indicate intelligence in space. Such signals have not yet been found, but it will be a long time before astronomers give up trying to figure out means of achieving communication with possible beings outside our solar system.

Winter Sky

Not far east of Cetus and reaching about the same meridian altitude, Orion rises in the evening in the late fall. This is a constellation no star lover in the northern hemisphere can miss. Orion is probably one reason why many persons think the stars

seem brighter in winter. In the summer sky there is no constellation to rival its brilliance. We met Orion in the late winter,
sinking into the west in the early evening and heralding the
approach of spring. When it rises in the east in the early evening,
winter is not far behind it.

Southern Hemisphere

A new set of constellations awaits the observer who journeys to
the southern hemisphere. The same axiom holds down there: the
latitude of the observer equals the altitude of the celestial pole;
but in this case it is the southern pole which is above the horizon
and the northern which has sunk down out of sight. There is a
southern circumpolar cap of constellations, just as there is a

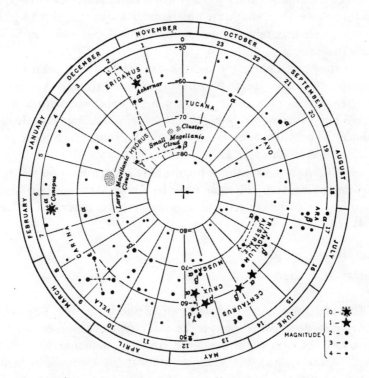

Figure 28. *The southern circumpolar constellations.*

northern, and again its radius is equal to the latitude of the observer. The farther south you go, the more the southern constellations are constantly above the horizon and fewer are the northern constellations that can ever be seen. Since we are writing primarily for the northern observer, we will mention just a few of the celestial sights you should not miss on a trip to the southern hemisphere.

The nearest star, and the third brightest, Alpha Centauri can be seen south of latitude 29° N. — the farther south the better. It and the bright Beta Centauri make a conspicuous pair east of the famous Southern Cross, which can be seen in its entirety from a degree or two even farther south. Parts of the Cross may be seen from as far north as latitude 34°.

There is no conspicuous star at the south celestial pole to help you find your way around, nor even a bright constellation. The south celestial pole lies in the feeble constellation of Octans. The nearest thing to a southern pole star is a 5th-magnitude star, Sigma Octans, barely visible to the unaided eye and almost 1° from the pole.

It is the masses of stars you can see in the southern hemisphere that are most important. Be sure to look in July and August at the Sagittarius star clouds passing high overhead with a brilliance they can never hope to achieve in the northern hemisphere. The Milky Way as seen from the southern hemisphere is described in Chapter 11.

Even though the southern hemisphere lacks a Pole Star, it has two magnificent circumpolar objects. These are the Magellanic Clouds, the two external galaxies nearest us, described in the final chapter.

Distances to Stars

With a telescope millions of stars can be photographed. But space is not crowded. Even though the stars are very numerous, the distances between them are large. In our particular neighborhood of the galaxy the stars average about three light years apart.

As we gaze out to the stars described in this chapter, we should realize that our solar system is like a little island in an enormous sea. The very nearest star, Alpha Centauri, is 4¹/₃ light years away. To get an idea of the chasm that separates us from the very nearest star, a small-scale model is helpful. Imagine the earth as a tiny ball ¹/₂ inch in diameter. On this scale the moon is a smaller ball, ¹/₈ inch in diameter and 15 inches away. The sun is a globe 4¹/₂ feet in diameter and 500 feet away. On this scale, where is the nearest star? It is 27,000 miles away!

★

★

★

NINE

Performing Stars

All that I know
  Of a certain star
Is, it can throw
  (Like an angled spar)
Now a dart of red,
  Now a dart of blue . . .

ROBERT BROWNING, "My Star"

There are stars in the sky that change their light by several fold or even many fold in the course of hours, days or months. These are called variable stars. They can also be called performing stars because they are putting on an act we can watch with interest. Some of them are bright enough that, without any telescope, we can watch them go through their gyrations.

  One obvious reason for a star's change in light is that

something — perhaps another star — at times gets in front of it and blocks some of its light from reaching our earth. This is happening to thousands of stars in the sky. In our solar system our sun is a single star. No second star ever blocks its light. We have a moon to block it, as we discussed in the chapter on eclipses.

Eclipsing Stars

A sizable proportion of the stars in the sky are actually double — two stars going around a common center of gravity. Frequently one of the pair is larger, cooler and darker than the other, and eclipses the bright member. The most famous such system is Algol, the Demon Star, also known as Beta Persei. It is normally a 2nd-magnitude star, the second brightest in the constellation and hence very easy to see. In mythology, Algol is the bright star in the head of Medusa, which Perseus is bearing.

This may have been the first variable star to be discovered. When the Arabian astronomers named it Algol in the middle ages they must have realized it was winking at them. The first written record of its variation, however, came from Montanari in 1669.

For a century no one seems to have tried to unravel its puzzle until a young Englishman, John Goodricke, solved the riddle when he was only 18 years old. Goodricke, a deaf mute who died at the age of 21, was favored with unusually keen eyesight. Also he had a close friend, Edward Piggott, whose father was an astronomer and who helped him night after night with the observations, in his garden.

On the evening of November 12, 1782, Goodricke watched Algol fade to only a third of its normal brightness. On December 28 again he saw it vary, and continued to observe it through the winter. By May 3, 1783, he had observed 11 passages of the star through its minimum light. He reported to the Royal Society that Algol varied in a period of 2 days, 20 hours, 45 minutes. After a year he revised this to 49 minutes, 8 seconds, an amazingly accurate value. Furthermore, he made the bold suggestion that the light variation could be accounted for "by the interposition of a large body revolving around Algol." This suggestion did not then meet ready acceptance, for as famed an observer as William Herschel stated that in his telescope he could see Algol only as a single star. Even today in our big telescopes Algol looks single. But a century after Goodricke, in 1889 in Potsdam, H. S. Vogel with a spectrograph proved that there are two stars in the Algol system. When measured with such an instrument Algol shows orbital motion around its center of gravity.

Goodricke's attention was not confined merely to Algol. On September 10, 1784, he discovered the variability of Beta Lyrae, an eclipsing star of a different type from Algol. Then a few nights later he found that Delta Cephei varied in a period of 5 days, 8 hours, 37½ minutes. This variation is not that of an eclipsing binary, but rather of a single pulsating star.

The work of Goodricke won great recognition during his brief lifetime. He was awarded the famous Copley Medal in 1783 when he was 19, and just two weeks before he died he was made a Fellow of the Royal Society. His death apparently was the result of long nights of cold, damp observing.

Ever since Goodricke's discovery, Algol has been intensively observed. Astronomers can spell out the details of the two stars in a rather amazing manner. The stars that are doing the eclipsing are much larger than our sun. The bright component, 150 times the sun's brightness, has three times the sun's diameter, and almost five times its mass. The fainter component has approx-

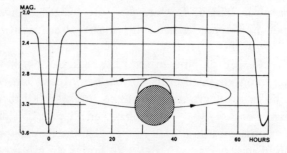

Figure 29. *The light curve of Algol, with diagram of the system. The larger, cooler star of the pair eclipses the brighter.*

imately the mass of the sun and a diameter 3.7 times as great. Though it is about 20 percent larger than the bright companion, it is three magnitudes fainter. The principal eclipse (primary minimum) therefore occurs when the large fainter star passes in front of the hotter, brighter companion. In between primary minima is a secondary minimum when the brighter star cuts off some of the light from the fainter. X-ray emission from Algol detected in 1975 indicates physical interactions between these two stars. Later researches have shown that the Algol system actually consists of three stars. The third has never been seen, but the motions of Algol indicate that there is a companion that goes around the eclipsing pair every 1.9 years at a distance about three times that of the earth from the sun.

Observing a Variable Star

Observation of variable stars can be very easy. That is, with no equipment except a simple star map you can estimate for yourself the magnitude of the variable and follow its changes from night to night. Simply compare the brightness of the variable with that of other stars whose magnitudes are shown on the map. Look back and forth from the variable to one of the marked comparison stars, and find the one it most closely resembles in brightness. If it falls between two, then estimate how much of the way from one to the other it seems to lie. You will probably be amazed at how expert you can become at this with a very small amount of practice. Professional astronomers love to go after variables with large telescopes and highly accurate photoelectric equipment. But this is not necessary to partake in the enjoyment of the star!

Eclipses of Algol come bunched at convenient times. They occur every three days, but come three hours earlier with each recurrence. This means that for about a third of the month the eclipses come at a time when it is convenient to watch them. This is followed by a longer period when they come in the daytime or at an inconvenient hour. For best viewing of the light changes, you should be able to see the star two or three hours before actual time of minimum. Watch it fade and then resume normal brightness. Predicted times of minimum for Algol are published in advance each year in *The Observer's Handbook*. It takes relatively little experience to become so familiar with the brightness of Algol and the surrounding stars that you can tell simply by looking at it whether or not Algol is undergoing an eclipse. It is a thrilling sight to be able to watch, just with your own two eyes, that mighty star, 100 light years away, fade until it is only a third of its normal brightness.

Another eclipsing binary is the star Beta Lyrae, whose light variations were also discovered by Goodricke. It is easily visible to the naked eye, and it, too, is the second-brightest star in its constellation, Lyra. Beta, a 3rd-magnitude star, is at a corner of the rhombus of Lyra, the long side away from Zeta. Sometimes Beta is brighter than Gamma, a short side of the rhombus away, and sometimes it is fainter. It varies from magnitude 3.4 to 4.3. Its period is 12 days, 22 hours, 22 minutes, but it is currently increasing at about 10 seconds a year.

The system of Beta Lyrae is much more complicated than that of Algol. It consists of a large blue star and a smaller white companion, which cuts off some of the light of the larger at primary eclipse. Puzzling features of the system showed up when spectra were obtained, and were very difficult to interpret.

Now, as a result of intensive work by many astronomers with modern powerful instruments, Beta Lyrae has yielded up some of its secrets. The two stars are elongated because of their strong mutual attraction, thus giving a complicated spectral pattern. Furthermore, as the stars revolve around their common center of gravity, great streams of gas issue from them and swirl around. No wonder this system bewildered astronomers for decades! But to the unaided eye the light changes are simple, easy to detect and fun to watch in the summer sky.

Intrinsic Variables

Beta Lyrae is a variable star in two ways. It is a geometric variable because, by the geometry of the situation, at times one of its stars prevents some of the light of the other from reaching earth. It is also an intrinsic variable because there is more change in light than can be explained from just the geometry of the system. The stars are actually physically interacting with each other, heating up and sending off gas streams. So Beta Lyrae serves as a good link between eclipsing stars and other classes of variables that are not geometric, but rather, intrinsic variables.

Most intrinsic variables have no second star that crosses in front of them and blocks some of their light from reaching earth. Rather, they are single stars going through a time of instability in their life cycle. The enormous amount of energy being generated in their intensely hot interiors is not flowing outward in a smooth and regular way, as it does for our sun and most of the stars. Instead, their radiation pours out by fits and starts.

There are several types of stars that vary their light intrinsically. One is the long-period variable. This type of star takes months or even years to go through its light cycle. And its change in brightness is very large indeed. A typical one may change its brightness by 5 magnitudes, a factor of 100; a rare one by 10 magnitudes, a factor of 10,000. The first such star to be recognized, and indeed the first star of which we have a written record of its variation, is Mira, the Wonderful Star, Omicron in the constellation of Cetus the Whale. On August 13, 1596 (before the invention of the telescope), David Fabricius, an astronomer of Emden and a friend of Kepler, recorded a 3rd-magnitude star in Cetus. Later, in October that year, he noted that it had disappeared. The star was assigned the letter Omicron by Bayer when he mapped and lettered the constellations in 1603. Actually, it was not until 1648 when Hevelius began a careful series of observations (which extended to 1662) that the periodic behavior of Mira become obvious. Although Fabricius has always been considered to be the discoverer of Mira, ancient Chinese observations recently compiled by Ho Peng Yoke state that a "Guest Star" was discovered in Korea on November 28, 1592 (Gregorian), and the probable identification of this star, which "diminished" on February 20, 1594, is that of Mira.

This giant red star has a diameter 300 times that of the sun. It overheats and blows off its outer veil covering. Then as it cools

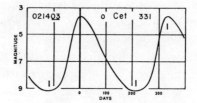

Figure 30. *An average light curve for Mira, omicron Ceti, period 331 days, as portrayed by Leon Campbell.*

down, the veil gradually settles back and the star fades away to obscurity for a while.

In some years Mira is as bright as 2nd magnitude. Then for three months its light decreases until the star disappears entirely to the naked eye. It can still be seen with optical aids. It remains

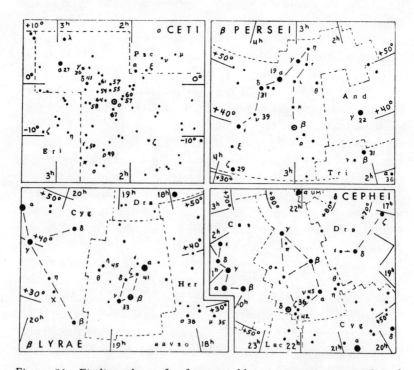

Figure 31. *Finding charts for four variable stars giving magnitudes of comparison stars.* (*From* The Observer's Handbook.)

below naked-eye visibility for about five months, and then gradually recovers its brightness in the next three. On average, the duration of naked-eye visibility for Mira in one cycle is about 18 weeks, but it may be as long as 21 weeks or as short as 12. Its period varies, as do the periods of most of the others in this class of long-period variables, by about 10 percent of its length. Mira reaches its greatest brightness in a period that is one month shorter than a calendar year. This means that for several years in a row the maxima come at a season when they can be readily observed. The maxima are now coming in the winter, when Cetus is well placed for observation. Mira came to maximum in February, 1976 and will come about a month earlier each succeeding year thereafter. The appearance of the constellation changes when this star suddenly pops up in it to naked-eye brilliance. Mira is but one of thousands of the type of long-period variable stars, but it is the brightest of its class, and by far the most spectacular for naked-eye observers. You find yourself asking if it is really believable that a great star, perhaps millions of miles in diameter, can actually change in such a fantastic manner!

Cepheid Variables

Another of Goodricke's variables, Delta in the circumpolar constellation of Cepheus, is a different type of intrinsic variable. Its circumpolar position means that the star can be seen from our northern latitudes on most nights of the year. Delta Cephei is somewhat less spectacular in its light variations than either Algol or Mira. Although its total variation in light is similar to that of Algol, from magnitude 4.1 to 5.2, it changes light more gradually, and takes 5 1/3 days to go through its light cycle. Its rise to maximum brightness is fairly steep, though not so steep as that of Algol coming out of eclipse, and its decline to minimum light is more gentle.

Delta Cephei is the prototype star for one of the most important classes of variable stars, the Cepheids, and was the first of the breed to be recognized. These stars act as celestial yardsticks. Cepheids are pulsating stars, going through an unstable phase in their evolutionary tracks. The length of the period of a Cepheid variable is directly related to its mass, which, in turn, is related to its absolute luminosity. This leads to a very important correlation, known as the period-luminosity relation. When the brightness of a Cepheid has been measured and its period determined, its distance is easily computed.

Fortunately, Cepheid variables are scattered throughout our galaxy, and the largest telescopes are able to photograph them in other galaxies to distances of millions of light years. The period-luminosity relation of the Cepheid variables has been one of the most important factors that have helped astronomers to measure the size of our galaxy and to determine the distances to the external galaxies.

Novae

The types of variables we have mentioned so far go on year in and year out repeating their light fluctuations over long periods of time. There are other types of variables, however, which appear and then after weeks or months during which they can be observed, they vanish, perhaps forever. One such type is the nova or new star. In a spot in the heavens where no star had ever been noticed before, suddenly one flares up. In a matter of days it may become as bright as any star in the sky. The star is not really a new one, but its increase of brightness by tens of thousands *is* new. In most cases, examination of photographs taken before the outburst reveals there had been a faint star — one of millions of faint stars — in that vicinity all along. Something caused the internal radiative equilibrium of the star to go awry, and the star literally blew its top. Shells of matter are hurled off with velocities up to several thousand miles a second. But actually only a small percentage, perhaps 1/10,000, of the star's mass is lost in the explosion, and the star is not extinguished by the experience. It may end as a white dwarf, a dying star whose energy sources are may become a white dwarf, a dying star whose energy sources are nearing exhaustion. Comparable in size to a planet, its density is millions of times that of water.

An outstanding example of such a star is Nova Persei 1901, discovered on February 22 of that year. It rose from fainter than magnitude 11 to brighter than magnitude 3 in 28 hours. For a while it was brighter than any star visible from mid-northern latitudes except Sirius. After a few months it faded below naked-eye brilliance, but some years later astronomers photographed the light of the explosion as reflected from clouds of gas surrounding the star. The diameter of this apparent gaseous envelope grew with time as the light traveled farther and farther out from the star.

Another very bright nova was Nova Aquilae in 1918. On June 5 it was of magnitude 11, as it had been for the 30 years that photographs of it existed. By June 8 and 9 it had reached magnitude −0.5, as bright as Altair. The following night it was even brighter, −1.1, and outshone every star in the sky except Sirius. In a few weeks it could not be seen without a telescope.

Sky observers had a rare treat on the last days of August, 1975, when a hitherto unrecorded star in Cygnus staged a massive flare-up to millions of times its original brightness. It was discovered first by Minoru Honda of Kurashiki, Japan, and reported by K. Osawa, Director of Tokyo Astronomical Observatory, to the Smithsonian Astrophysical Observatory. With the greatest rise on record for any nova, more than 19 magnitudes, it was brighter than magnitude 2.0 at maximum, and was independently discovered by hundreds of observers. For only a week it lingered at naked-eye brightness, falling about 1 magnitude a day after maximum. Then the rate of decline slowed. It can probably be reached with telescopes for many years to come. In characteristics it had some resemblances to a supernova, discussed on page 201.

Every century, dozens of such new stars appear. Estimates are that as many as 25 novae may occur in our galaxy every year. Most of these are too faint to be found, even at their greatest brightness, and some are completely hidden behind clouds of obscuring matter. Some are not found until years after their outburst, when photographic plates are compared. Keeping track of sporadic events in the 41,253 square degrees in the sky is a task that can be accomplished only by thousands of observers.

Accordingly, many amateur astronomical societies have "Nova Search" programs, and many individuals with binoculars have trained themselves to recognize stars below naked-eye brightness in various regions of the heavens. A telescope is not necessary for such a search, but binoculars are a great advantage.

For one very assiduous observer, G. E. D. Alcock of England, such training and diligence paid off handsomely. In 1955, Alcock began the tedious task of learning the appearance of star fields in binoculars. On January 1, 1961, he began systematic searching for changes in these fields. After 800 hours spent in such searches his diligence reaped a rich reward. There, on July 8, 1967, in a crowded Milky Way field in Delphinus was a new star, barely of naked-eye visibility, magnitude 5.6. This nova did not reach the great brilliance of Nova Persei or Nova Aquilae, but it did reach 3rd magnitude, and stayed at naked-eye visibility for many months. It proved to be a very remarkable "slow" nova, which gave astronomers an unusual opportunity to study the behavior of such a star.

After an observer thinks he has detected an important change in the sky, there is a flurry of activity to get the word to the astronomical community around the world. Such a broadcasting of information is done from the Smithsonian Astrophysical Observatory, Cambridge, Massachusetts, where the Central Bureau of Astronomical Telegrams is located. The happy discoverer sends a telegram or cable to this Central Bureau, usually after his discovery has been confirmed by another observer. The Smithsonian asks other observers to make an independent check. (In the case of Nova Delphini 1967, Alcock's discovery was confirmed in England by Michael Candy of the British Astronomical Association. Several hours later the message was at the Smithsonian, where verification of discovery was obtained before the world was alerted.)

The discovery message then goes by telegram to 30 observatories in North America, and others in France, Brazil, Chile, Japan, Australia, South Africa and Argentina, with a distribution chain of 65 observatories in Europe and Asia. The telegram is composed in a compact coded numerical form, which provides an internal check that the important positions and numbers have been relayed correctly. Sometimes, if the object is sufficiently important, phone calls are made to certain observatories.

In addition to the observatories receiving telegrams, there is a postcard service, and 600 subscribers receive their cards several days after the telegrams are sent. These cards are preaddressed by computer, then printed on a rush basis by photo offset reproduction. Thus, a vast observational network links the astronomers of the world.

Supernovae

Once in several centuries in a galaxy a really great star outburst occurs — a supernova. Whereas an ordinary nova may rise to an absolute brightness 100,000 times that of our sun, a supernova goes up to a brightness of tens to hundreds of millions of suns. A supernova may give off in one day as much radiant energy as our sun gives off in a million years. A supernova is the greatest catastrophe known in the universe. A large percentage of the mass of the star is destroyed as it is converted to energy. Whereas an ordinary nova continues on as a regular star, the remnants of a supernova might be a neutron star, at the last stages of its existence, and expanding shells of gas. The cause of a supernova explosion is not yet clear. It is accepted, however, that the enormously high temperatures of the supernova holocaust lead to the formation of chemical elements of higher atomic number.

Such a star was seen by the famous medieval astronomer Tycho Brahe on November 11, 1572, in the constellation of Cassiopeia. While on his way home from his chemistry laboratory, Tycho saw the brilliant star in a portion of the heavens where he knew none existed. The story goes that he was so astonished that he asked the servants accompanying him if they, too, could see a bright star in that spot in Cassiopeia. They certainly could! When Tycho first saw it, it was already brighter than the planet Jupiter, near magnitude −2. In a day or two it became as bright as the planet Venus, and hence visible in the daytime.

Many observers all over Europe saw the star at the same time, and Tycho is not credited with being the first to discover it. But he wrote a very important treatise on it. By observing the star at 12-hour intervals, above and below the pole, he determined that it was farther from earth than the moon is. In keeping with his role as a nobleman, he was reluctant to publish his paper, but his friends persuaded him to do so, and it became the first of many. The influence of the star on Tycho was so great that it diverted him from chemistry, which he had been pursuing, and propelled him into a lifetime of astronomical work. He became an outstanding astronomer after the King of Denmark set him up on the island of Hven, between Denmark and Sweden, with a splendid observatory. Since this was before the invention of the telescope, the instruments were sextants, quadrants and astrolabes, all without optics. It was a feat in itself to lift these instruments, so massive and cumbersome were they.

The supernova of 1572 was visible to the unaided eye for 17 months. It faded gradually and changed color, and then was lost. In the present century, astronomers have actually photographed some of the faint gaseous material thrown off when the star exploded. Just as the Star of Bethlehem heralded the birth of Christ and the Christian era, so the nova of 1572 was supposed to herald another era. To the pessimistic it betokened the end of the world.

The story of the supernova of 1054 A.D. in Taurus is even more exciting, and more fully developed. Its tale has unfolded through several decades of modern astronomy. For the Western World, the story begins in the 18th century. At that time a famous French astronomer, Charles Messier, was diligently hunting for comets. He found that he was frequently mistaking for comets certain luminous fuzzy patches in the sky. Accordingly, he determined to make a list of these objects, with their accurate positions, so that they would not forever be getting in his way and tricking him. With his list he hit the jackpot of approximately the hundred most important objects in the sky outside our own solar system, visible from his latitude!

Number one in his list is a fuzzy patch in the constellation of Taurus, which later became known as the Crab Nebula because of its shape. Actually, this had already been discovered in 1731 by John Bevis, an English physician and amateur astronomer, but the publishing house to which he entrusted his work for publication went bankrupt, and not until 1786 was his atlas published. Meanwhile Messier had independently discovered the Crab in 1758 while following the comet of that year.

During the 19th century the luminous patches numbered by Messier became much better understood. Some are luminous masses of gas; others are clusters of stars so far away that the individual stars could not then be resolved; still others are whole galaxies of stars, at distances of millions of light years. For many decades Messier 1 stood out by itself. It had a more symmetrical structure than most of the so-called diffuse nebulae. Why? In the 1930's velocity measures with the Mount Wilson telescopes showed that there was an outward expansion of the nebula from the center with a velocity of 680 miles/sec. or 60 million miles a day. At the accepted distance of the Crab, 6,000 light years, this indicated that the expansion had started between 1000 and 1100 A.D. About the same time as the Mount Wilson measures were made, a search of Chinese records revealed that the Chinese had recorded a "Guest Star" in the constellation of Taurus in the year 1054 A.D. This star was as bright as Jupiter and was visible for two years.

Now it is suggested that the Indians of the southwestern United States also saw and recorded this supernova in pictographs. A symbol of a crescent moon with a circle just below it found in Indian caves and previously unexplained may refer to this outburst. One was cut into a Navaho canyon wall, and another was drawn with a lump of hematite (a red ore containing iron) on a cave wall at White Mesa, Arizona. As the crescent is not common among the petroglyphs and pictographs of northern Arizona, it baffled the experts until the association with the supernova was suggested. Dr. William C. Miller of Mount Wilson and Palomar observatories says that early on the morning of July 5, 1054, before dawn the crescent moon stood just 2° north of the supernova. This was a spectacular configuration. The supernova was the brightest star (other than the sun) that had ever appeared in the sky within the memory of man. For 23 days it was bright enough to be visible in daylight. It could be seen at night for almost two years. These two prehistorical cliff sites had an unobstructed view of the eastern sky, and the caves were occupied at the time the supernova burst forth.

But the saga of the Crab Nebula does not end here. For when radio astronomy came into being, and radio telescopes began to sweep the heavens, collecting long-wave radiation, the Crab Nebula proved to be one of the strongest sources of long-wave radiation picked up by radio telescopes in the sky. It now has another designation, the type given to strong radio sources, Taurus A. And in the fall of 1968, the year pulsars were discovered (see Chapter 10), a pulsar was found in the Crab. A pulsar is a neutron star a few tens of miles in diameter with a density a million million times that of water. This one is spinning all the way around on its axis in .033 second of time. It may be the residual cinder of the violent explosion witnessed from earth in 1054 A.D.

So Messier 1 is the Crab Nebula, is the Supernova of 1054 A.D., is Taurus A, and contains Pulsar NP 0532.

Galactic Supernovae

There are now five recognized supernovae that occurred in our galaxy in the second millennium A.D. The two just described in detail are the most thoroughly investigated. The others are one in Lupus in 1006, Kepler's Nova of 1604 and that which produced the strong radio source known as Cas A, about 1667. Dr. Sidney van den Bergh estimates that about 20 supernovae occur in our galaxy every millennium, a rate comparable to that in other galaxies of the same spiral type as ours. Only every few hundred years, however, is one sufficiently bright to be spectacular. And many have been recorded in other galaxies, but these even at their brightest are below naked-eye visibility because of the great distances involved.

Many events in astronomy are predictable — eclipses, some meteor showers and light variations of many stars. It is unlikely that in the foreseeable future the outbursts of novae or supernovae for any of the 200,000 million stars in our galaxy can be predicted in advance. The only way we can catch these outbursts and learn more about their cause and about the stars affected by them is through a constant vigil of the heavens. The largest telescopes in the world are not suitable instruments for such a search, because through them astronomers can see only a tiny portion of the heavens at one time. Accordingly, this is a field for search with the naked eye or with binoculars, something that can be a happy pastime for amateur astronomers. Imagine the thrill it would give you to discover a "new star," never before seen by human eyes! And at any time, with no advance warning, a new supernova could burst forth in our galaxy and become the brightest star in the sky.

Amateur Astronomers' Organizations

In 1911 an organization known as the American Association of Variable Star Observers (AAVSO for short) was formed by amateur sky watchers who were taking delight in making observations of variable stars. This was an outgrowth of an urging by the famous German astronomer, F. W. Argelander. Decades earlier, he had sensed that there was more going on in the sky than the limited number of professional astronomers could keep track of. Even today, with the vast growth in the numbers of astronomers and the capabilities of their instruments, this still holds true.

The AAVSO has been flourishing ever since, and now has nearly a thousand members in many countries of the world. The total number of observations of variable stars made by members of this organization, mostly with telescopes, has now passed the 3½-million mark.

A foremost amateur in North America is one of the AAVSO's prime observers, Leslie Pelletier of Delphos, Ohio. Mr. Pelletier's fabulous record is that he has made more than 127,000 observations of variable stars and discovered 12 comets. Popular books on astronomy first sparked his interest. He has described his astronomical yearnings, delights and successes in an engaging autobiography, *Starlight Nights*.

The AAVSO is but one of many organizations of amateur astronomers. There is a unity of purpose among the members of all of them: to enjoy the wonders of the skies to the full. Many of the members work energetically to increase our knowledge of the heavenly bodies. They make their own telescopes, many of them of high quality. They spend endless hours estimating the brightness of variable stars, searching the sky for new stars to appear, hunting for comets or recording their appearance, tracking meteors, counting sunspots, classifying auroral appearances into their various forms such as draperies and arcs, and precisely timing occultations of stars and planets by the moon. Many become expert and useful photographers of the sky. While pursuing their vocation by day, they devote themselves to their hobby at night.

There are other kinds of members of the organizations, too, the "Armchair Astronomers." They lend valuable support with interest, membership and donations, but are content to absorb their astronomy at meetings or at home with books and periodicals. Not for them are the stiff necks, the chattering teeth and the icy fingers from stargazing on bitter winter nights. But they bask in the enjoyment of the knowledge of the universe acquired by others.

In Canada, the Royal Astronomical Society of Canada has been in existence since 1890. It has nearly 3,000 members in 18 Centres, from St. John's, Newfoundland, to Victoria, British Columbia, or as members at large in a dozen other countries. The Centres hold regular meetings, sometimes with distinguished astronomers speaking in popular vein, and sometimes with members discussing their observations. The Society publishes a bimonthly *Journal* with articles of interest, and annually *The Observer's Handbook*. This is invaluable to anyone who wants to follow celestial happenings, such as positions of the bright planets, phases of the moon, and so forth.

In the United States, the Astronomical League, established in 1947, acts to coordinate amateur astronomers. It has 156 member societies, ranging in size from 4 to 325 members, totaling about 7,500 amateur astronomers. The AAVSO and the Association of Lunar and Planetary Observers are affiliated with it. The League holds both national and regional conventions and publishes a quarterly journal, *The Reflector.*

The International Union of Amateur Astronomers held its inaugural Congress in Bologna, Italy, in 1969 with representatives from 16 countries. Its purpose is to coordinate the activities of amateur astronomers throughout the world. Every three years it holds a General Assembly, the last being in Hamilton, Ontario, in July, 1975.

In the British Isles, the British Astronomical Association has been carrying on valuable work since 1890, with different observational sections for planets, meteors and so on. Regular meetings are held and a monthly journal and annual handbook are published. Another long-established organization is in France, the Société Astronomique de France, formed in 1887. Its monthly bulletin appears as the beautifully illustrated periodical *L'Astronomie,* a monthly review now in its 89th year, founded by the incomparable popular writer of astronomy, Camille Flammarion.

Scattered around the world, including the southern hemisphere, are many other similar societies. Many amateur astronomers are members of more than one society and send their observations to the most appropriate organization.

★

★

★

TEN

As The Heavens Unfold

When night hath set her silver lamp on high
Then is the time for study.

PHILIP JAMES BAILEY, FESTUS, SC.,
A Village Feast

Up to this point we have been talking almost entirely about the objects that man has seen in the sky for thousands of years. And even in early times a few brilliant souls had some understanding of what they were looking at — the sun, moon and planets.

Now we come to the distant branch of astronomy whose understanding belongs to the 20th century. Not yet four centuries have elapsed since man first gazed at the skies with optical aids to supplement his own eyes. Nowadays in North America there are thousands of small telescopes, and binoculars are a common asset of many households. It is difficult for us to imagine now the sparse knowledge of the skies in the millennia when lenses did not exist. With the naked eye, the world over you can see only about 6,000 stars. With a three-inch telescope in your hand you could see more than 100,000 stars. And with the largest telescopes existing today probably a billion could be photographed.

The early stargazers did have one great advantage, however — dark skies. Only the moon and an occasional aurora diminished the brilliance of the stars and planets for them. But lacking were details of the objects they saw — their eyes were not adequate. Before the 17th century, the instruments used to study the sky were merely to measure positions of the heavenly bodies. They could not gather more light from them, or magnify them. Even in the late 16th century when Tycho Brahe heaved his sextants and quadrants around in his observatory he could not *see* the planets any better than we with our own eyes — he could only measure their positions.

Telescopes

The great leap forward for astronomy came in the first decade of the 17th century when a Dutch spectacle-maker, Jan Lippershey, invented lenses that would form a magnified image of a distant object. And the Italian physicist-astronomer, Galileo Galilei, hearing of the invention, was able to fashion such lenses himself and turn them to the skies. There he saw the craters and valleys of the moon with a clarity hitherto undreamed of. He saw four

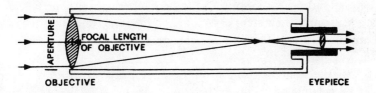

Figure 32. *In a refracting telescope, starlight is brought to a focus by passing through the objective, a lens.*

tiny "stars," which proved to be moons accompanying the planet Jupiter. And the clouds of the Milky Way, known already for countless centuries, he saw dissolve into myriads of stars.

The earliest telescopes were refractors. The object glass that gathers the light and projects the image was a lens. (The light-gathering power of a telescope is proportional to the area of the aperture, i.e., the square of the radius. The magnifying power is directly proportional to the focal length.)

Since the first telescope was built, astronomers and other scientists have been engaged in a constant improvement and enlargement of telescopes and their auxiliary equipment. Half a century later the great Sir Isaac Newton perceived that, as the object glass for a telescope, a mirror would have certain advantages over a lens. The reflecting telescope was born. Gradually both refractors and reflectors grew larger and larger.

The 19th century saw a summit reached in the size of the refracting telescope. The largest in the world is the 40-inch refractor at the Yerkes Observatory of the University of Chicago at Williams Bay, Wisconsin. A close second is the 36-inch refractor of the Lick Observatory of the University of California, atop Mount Hamilton not far from San Francisco. At the time of the great San Francisco fire, astronomers at Lick with this instrument had an unforgettable view of the conflagration, which seemed like the flames of hell on earth.

The 19th century also saw the development of the spectro-scope, an instrument that breaks light up into its constituent wavelengths by a prism or a grating. The spectroscope brought the knowledge that the same chemical elements exist throughout the visible universe — though the proportions in different bodies do differ. Not merely can the composition of the stars and nebulae be measured, but also their temperatures, densities and pressures and their sizes determined. It is sometimes said that we know more about the interiors of some distant stars than we do of our own earth!

And the same century brought photography, without which our knowledge of the universe would be miniscule. Spectroscopes known as spectrographs were developed to take photographic plates. These are invaluable tools for optical telescopes.

With the 20th century came great increases in the size of reflecting telescopes. In 1918 the 100-inch reflector of the Mount Wilson Observatory of the Carnegie Institution of Washington began operation. It was the power of that telescope which first resolved the distant "island universes," the external galaxies such as the Great Nebula in Andromeda, into stars. This made a gigantic leap in man's knowledge of the universe.

Then in 1948 the great 200-inch Hale telescope began operation on Mount Palomar, as part of the Mount Wilson and Palomar (now the Hale) observatories. Observations with this instrument gave Dr. Walter Baade the clues to a better distance scale of the universe than we had previously.

For one brief period the largest telescope in the world was in Canada. For a few months in 1918 the 72-inch reflecting telescope of the Dominion Astrophysical Observatory on Little Saanich Mountain, near Victoria, B.C., held that honor, before the 100-inch went into operation.

For about two decades Canada had the second largest telescope. The DAO instrument held that place until May, 1935, when the 74-inch reflector of the David Dunlap Observatory of the University of Toronto began its work at Richmond Hill, Ontario. It retained second place until 1939, then was surpassed by the 82-inch reflector of the McDonald Observatory in Texas.

The world's largest optical telescope now is the Zelenchukskaya, a giant 6-meter in the Caucasus Mountains of the Soviet Union overlooking the Bolshoi Zelenchuk gorge. Started in 1960 as a project of the Soviet Academy of Sciences, its 25,000 parts have taken many years to assemble and adjust. The mirror of low-expansion borosilicate glass weighs 42 tons — the whole telescope, 850 tons.

An Astronomer's Night and Day

The public image of an astronomer is of one who works with a large optical telescope. Such an astronomer's night goes like this, when it is clear enough to work. As daylight fades, the observer swings open the shutters of the dome. The telescope has been left

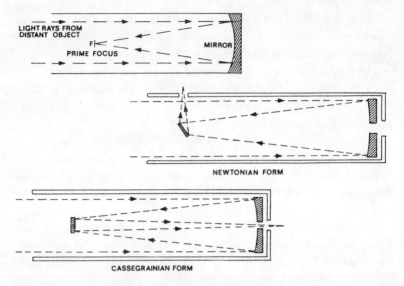

Figure 33. *In a reflecting telescope, starlight is brought to a focus by reflection from a mirror as the objective. Top: Prime focus. Middle: Newtonian. Bottom: Cassegrain.*

in its rest position, with the tube horizontal. The observer mounts to the end of the large mirror and rolls back the dust cover from the great glass and, at the other end of the tube, uncaps the secondary mirror. This may be the Cassegrain type, which reflects the light back down the tube through a hole in the main mirror. Under the big mirror are eyepieces and holders for photographic plates to be used with the Cassegrain mirror. If the secondary mirror is a flat, Newtonian type it throws the light off to the upper edge of the tube. The astronomer must be able to reach this top edge, sometimes high up in the dome, to place the plateholder on the side of the tube, right at the top. There the plates are exposed. Some of the more recent large telescopes now are built with a cage at their upper end where an observer may work in the center of the large tube at what is called the prime focus.

Sometime during the getting-ready process the astronomer throws a switch to open the shutters of the dome. As twilight fades, the astronomer prepares to set the telescope on the desired object. You must know star time, sidereal time. (This is technically the right ascension of the meridian at the given instant.) Near the telescopes, observatories keep both sidereal and standard time clocks. Then from a catalogue or observing list the observer reads the two essential coordinates, right ascension and declination, of the object to be studied and sets the telescope in these. After the glass photographic plate is put in the plate holder (most professional astronomical photography is on glass) and the sky is dark enough, the exposure may be started. Depending on the brightness of the object under study, exposures may run from seconds to full nights, or in rare instances, many nights. The telescope is driven by controls to follow the object, but the observer must check on it from time to time — a process called "guiding." The dome must be kept at outside temperature so that warm currents will not cause jiggling of the image. Observing in the long winter nights in severe climates becomes a challenge! Some astronomers take to heated flying suits and special heavy winter boots.

Once the plates have been taken, they must be developed — usually the next day. The size of the plate is almost in inverse ratio to the size of the telescope. For the big reflectors, the plates are only a few inches in diameter, square or rectangular. Spectrum plates may be as small as $3\frac{1}{2}$ x 2 inches. A little plate like that sometimes holds the results of several nights of work with the most valuable and sophisticated astronomical equipment. It takes strong nerves on the part of the observer to develop an investment like that, and see the result!

An astronomer will then spend weeks or months measuring the plates acquired. A variety of machines can measure the size and blackness of the star images, or make precise measurements of positions of stars and other celestial objects. Sometimes one night's work means more than a month of office work to analyze the results. Very few astronomers spend more time at night observing with a telescope than they spend in their offices by day, deriving the results of their nighttime labors and preparing them for publication.

Not all astronomers nowadays take photographic plates when they observe. Nor do they work instead, as did astronomers of old, with an eye close to the eyepiece of the telescope. Scientists in this century have been very successful in developing auxiliary equipment that actually increases the power of the telescope on which it is used. The photoelectric image tube and image intensifier are now widely used.

The image tube, made of material that releases electrons when light hits it, receives at one end the light gathered by a telescope. This incoming light then ejects an electron, which falls down the tube, propelled by a strong electric field. When it reaches the other end of the tube it makes a phosphor glow in a phosphorescent screen. The magic part is that the electron has picked up energy during its fall and makes the phosphor glow brighter than at the start. This gain in brightness enables the image tube to reach fainter objects than the photographic plate. Up to three or even four image tubes have been linked together to produce a tremendous light amplification. A gain of 10 to 20 times may be achieved.

The technique can be improved further by adding a TV camera, which experts say is more efficient than the eye for acquisition of faint objects. Then a minicomputer is used to control the instrument and process the data. The astronomer can now operate the equipment in a comfortably air-conditioned room, where he can work most efficiently. In the room with the computer and TV camera controls and monitor is a console that contains all the telescope and dome controls. More than 40 percent of the time of the Kitt Peak National Observatory 2.1-meter telescope (near Tucson, Arizona) is now scheduled for image-tube work.

The apparatus is so delicate that objects too faint actually to be seen even on the TV screen can be studied. This is done by offsetting the photometer a precise, known distance from an object that *is* visible. The central region of one of the nearby faint galaxies, the dwarf spheroidal Andromeda I has been studied in this way.

The last third of the 20th century has brought increased effort to ensure not merely that larger telescopes are built, but that they are placed in the most favorable atmospheric and climatic positions possible. For ground-based astronomy, this leads to earthly sites far away from the sponsoring institution, sites where cloudy nights are few on an annual basis, and where a remarkably steady atmosphere keeps twinkling of stars to a minimum.

Certain mountaintops in Chile, between the Pacific coast of that country and the high Andes farther inland, have been found extraordinarily fine for these conditions. Consequently, a great concentration of astronomical equipment is being placed on these peaks, despite the politically unstable climate of that country. (All factions in Chile look kindly on astronomical work, which removes no natural resources from the country.)

Kitt Peak National Observatory of the United States sponsors the Inter-American Observatory at Cerro Tololo, with headquarters in the old port city of La Serena, 60 miles west. In 1974 the size of telescopes in the southern hemisphere began to approach that in the north when a 158-inch reflector began operation on Cerro Tololo, where seven smaller telescopes were already in use.

The Carnegie Institution of Washington has its southern observatory (CARSO) on Cerro Las Campanas, 100 miles north of its headquarters in La Serena. CARSO has a 40-inch reflector in operation and in 1976 may come the completion of its 100-inch DuPont telescope. By arrangement, the University of Toronto has its own 24-inch reflector on the same mountain, with an elevation of over 7,000 feet. The scale of the University 24-inch telescope at its Cassegrain focus is practically the same as that of the David Dunlap 74-inch at its Newtonian focus, where direct photographs are taken.

Six European countries have banded together to establish a fine European Southern Observatory (ESO) on La Silla, a mountain close to Las Campanas.

In 1974 the Anglo-Australian telescope was inaugurated by H.R.H. the Prince of Wales on October 16 at Siding Spring Observatory near Coonabarabran, New South Wales. This 3.9-meter reflector (150 inches) is jointly funded by the British and the Australian governments. This giant, along with the new 158-inch on Cerro Tololo, will reveal parts of the universe hitherto unattainable. The 74-inch reflector of the Mt. Stromlo Observatory near Canberra has probed the southern skies for many years.

The most exciting project for Canadian astronomy in many years is the CFHT, the Canada-France-Hawaii telescope to be placed on Mauna Kea, Hawaii. With an aperture of 3.58 meters (144 inches), the time of the instrument will be shared 42.5 percent by Canada and France, and 15 percent by the University of Hawaii, which is responsible for roads and some buildings. The Cer-Vit mirror is being ground and figured in the optical shop of the Dominion Astrophysical Observatory at Victoria, B.C. It is hoped the instrument will be operational in the last years of this decade. Other major telescopes are already in operation in Hawaii.

Radio Telescopes

Besides being an era of major development in optical telescopes, the 20th century has seen the discovery of the radio telescope, an instrument that attains a deeper study of the heavens. Radio telescopes do not displace optical telescopes; they supplement them, helping to bring us a fuller, more balanced picture of the universe.

A radio telescope has a large antenna that collects from celestial bodies the radiation that is too long to be seen in optical telescopes. This radiation ranges in length from a few millimeters up to many meters. By good fortune a window in the earth's atmosphere permits entry of these wavelengths. In 1932, Karl Jansky of the Bell Telephone Laboratories in New Jersey was working with a large radio receiver to investigate static. He recorded strange sounds coming from the heavens. From the times of its appearance, he proved that it was originating from the starry heavens. And with that discovery, supplemented several years later with work by Grote Reber, the great field of radio astronomy was born.

Now giant ears are dotted all over the world. Early philosophers would say they are listening for the music of the spheres. Some radio telescopes are single large parabolic reflectors. The best known of these probably is at Jodrell Bank, where the 250-foot dish was the only large telescope in the non-

Communist world able to track the first Sputniks. Canada has a high-precision 150-foot dish at the Algonquin Radio Observatory in the well-known provincial park of the same name. This telescope was used for a great scientific breakthrough when it was teamed up with the smaller 85-foot dish of the Dominion Astrophysical Observatory at Penticton, B.C., to make the LBI — long baseline interferometer. This enabled positions of radio sources to be measured with a far higher degree of precision than ever before. The concept has been further developed on an international scale, linking telescopes at great distances, in Australia, the Soviet Union, Britain, the United States and Canada. Ultimately a linkup with a telescope in space is possible. The largest single dish is the 1,000-foot at Arecibo, Puerto Rico. This new dish had a multimillion dollar face-lifting, completed in 1974, with a new surface that greatly increases the capability of the instrument in the shorter wavelengths.

Then there are arrays. These may consist of several or many medium-size dishes linked together to give a much greater power. For example, an array of a type known as an aperture synthesis telescope devised by Nobelist Sir Martin Ryle consists of eight telescopes placed along a railway track and used to build up a picture that would be given by a single telescope 3 miles in diameter. Or arrays may be a series of poles with wires strung on them. A T-shaped array at the Dominion Radio Astrophysical Observatory has a collecting area of 17 acres. Its east-west arm is 4,224 feet long, and the north-south 7,430, with 624 wire-strung tall poles acting as "full wave dipoles."

An outstanding array is under construction for the National Radio Astronomy Observatory, which for many years has had a wealth of equipment in a quiet valley at Green Bank, West Virginia. The radio telescopes there include a 300-foot dish, a 140-foot dish of high precision and three 85-foot dishes mounted on a roadway for use as an interferometer. The new project, known as the VLA for Very Large Array, is in a remote part of New Mexico with particularly suitable properties, the Plains of San Augustin, 7,000 feet above sea level and some 50 miles west of Socorro. The array will consist of 27 solid surface reflectors 82 feet (25 meters) in diameter mounted on an equiangular Y-shaped configuration. This will give the equivalent power of a single radio telescope antenna 17 miles in diameter! The instrument is expected to reach its full capacity in 1981, but becomes operational in 1976 when six reflectors are in place.

The revelations of radio telescopes have yielded one surprise after another to astronomers. For example, they have shown the planet Jupiter as a 100-million-watt natural radio transmitter. They have shown the existence of vast clouds of gas in our galaxy. Neutral hydrogen was expected, but now more than 30 molecules, some quite complex, have been identified. One of the latest, No. 30, in the beautiful and important Orion nebula cloud, is a nine-atom molecule, dimethyl ether, consisting of two atoms of carbon, six of hydrogen and one of oxygen. Some of these molecules have characteristics related to building-blocks for life.

With the radio telescopes, hitherto unknown and unsuspected types of objects have been revealed. First to be found in the early 1960's were the quasars, discussed in Chapter 13.

Then in February, 1968, the discovery of pulsars was brought to an astonished world. Anthony Hewish and graduate student Jocelyn Bell detected them with the Mullard radio telescope of the University of Cambridge with an array described as 4½ acres of washing line. They obtained 3½ miles of recorded data to verify the incredible objects.

It is no secret that, when pulsars were discovered, the reputable scientists who made the discovery held back on the announcement for some weeks because they wondered if these incredible signals were actually from the LGM. "Little Green Men" is a favorite phrase for the beings that might inhabit outer space. It soon became obvious to the scientists that with more than one pulsar, and with the tremendous amount of energy involved, it was impossible to ascribe the pulsars to anything but a natural phenomenon. Eventually, though, some radio telescope may pick up a pattern of signals that is being sent by an intelligent being.

Only a few months of observations were needed before astronomers decided just what pulsars must be. They are stars that are spinning on their axes in a fraction of a second, one (the

Crab pulsar) in as short a time as .033 seconds, and doing this with the utmost regularity. As they spin, a high-energy spot on the star gets beamed toward earth, giving a pulse of radio energy that we receive with great regularity. The pulsars must be neutron stars — no other explanation can suffice. A pulsar has a mass similar to that of our sun, but its diameter is only a few miles. It is the end product of a star — perhaps of a supernova explosion like that of the Crab Nebula — in which the atoms are stripped of their electrons and the nuclei are packed so closely together that the density amounts to 100 million million times that of water. A cubic inch of the stuff would weigh one billion tons! And the electrical energy from ¼ square inch of the surface equals the entire electrical output of the earth. At present about 120 pulsars have been identified.

One further step toward an exotic body could be a black hole. This catchy phrase has captured the public imagination. Black holes were theorized in the 1930's, but only recently have observational astronomers become interested in them. A black hole is considered to be a region of space into which a star or massive collection of material has fallen, and from which no light or matter or signal can escape. After matter has fallen into a black hole it has no distinguishing characteristics left. Only mass, charge and angular momentum remain. Now, astronomers working with evolutionary theories have long realized that if a star is too massive, it cannot subside into a white dwarf, or even into a neutron star. Some stars have a mass 50 times that of our sun. The gravitational effects in the star's collapse would dominate, and the matter would pass beyond the density of a neutron star into that of a black hole. The black hole has a spherical surface, with the radius proportional to the mass in the hole. If the sun collapsed down to a black hole, it would be only four miles in diameter. It is not yet known how much mass a star must have to become a black hole and not a neutron star — perhaps three times the sun's mass is enough.

Astronomers are now greatly intrigued to study the sky for evidence of possible black holes. These might be detected through their gravitational influence on another body. A double-star system that has a massive invisible component with peculiar characteristics is a good candidate. Such a one is the X-ray source Cygnus X-1, which is associated with a binary star system called HDE 226868. (HDE stands for the Harvard Catalogue known as Henry Draper Extension.) The system is 8,000 light years or more distant. The visible star has a mass 30 times that of the sun, the small dark companion has 10 times the sun's mass.

Theory has it that matter would swirl into the black hole from the nearby companion, giving off great energy in the form of X-rays. The Copernicus satellite has detected a period of 5.6 days for Cyg X-1, a period also found in spectrographic work by Dr. Tom Bolton with the David Dunlap 74-inch reflector. These observations help to determine the masses of the two bodies and corroborate the idea that the unseen one is a black hole.

Communication with Extraterrestrial Life

In the minds of most astronomers there is little doubt that life exists elsewhere in the universe. For one thing, scientists shy away from the anthropocentric viewpoint that our earth and the beings on it are unique in this vast and massive universe. Although they have not been actually seen, the probabilities are that there are thousands of millions of planets in our galaxy. Many must have conditions as favorable to life as our own. Then, too, the recent discoveries of great cosmic clouds of molecules that could be built up into living forms reinforces the idea that materials are available for life beyond our earth in deep space. The form that life might take is still a controversial subject. A considerable weight of opinion is that it would be oxygen-based, like our own, and not based on some other common element like silicon.

If we admit the postulate that there is life elsewhere, it follows naturally that such life may have developed a civilization with which we can communicate. In fact, according to the experts, there is a good chance that such a civilization could be much more advanced than ours. If we get into communication with them, from a greater depth of experience they could teach us much from which we could profit — how to avoid nuclear wars, for example.

Radio telescopes seem to offer the best tool for such communication, though there are studies involving light concentrations such as in laser beams for sending signals. The large radio telescopes could now receive signals sent by a civilization as much as hundreds of light years from us — far, far beyond our solar system. Equipped with radar, they can send signals to distances of thousands of light years.

The first systematic listening for extraterrestrial artificial signals was by Dr. Frank Drake in Project Ozma in 1960 with the 85-foot telescope of the National Radio Astronomy Observatory at Green Bank, West Virginia. Drake selected several nearby stars of the right type possible to have planets. The experiment revealed no signals, but later theoretical work showed that on the basis of probability, a distance of only ten or so light years was too small to bring success. However, by the time you try to go to the hundreds of light years necessary for success you have a major problem. From the thousands upon thousands of objects available, how do you pick the right ones on which to concentrate the valuable telescope time?

Meanwhile we *are* trying to call attention to ourselves far out in the depths of space. Pioneer 10 was equipped with a specially designed plaque intended to convey information about the earth, earth dwellers and the solar system pictorially. After leaving Jupiter it headed for the distant spaces between the stars where perhaps some other civilization can capture it.

And when the giant Arecibo 1000-foot dish was dedicated in November, 1974, after its resurfacing, a message in binary code conveying vital data about the solar system and the earth people was beamed with the 450,000-megawatt transmitter, using the great dish, to the globular cluster Messier 13 in Hercules, discussed in Chapter 12. Planetary expert Carl Sagan thinks there is one chance in two that there is a civilization within the cluster that could receive this message. In a gamble of this nature lies one comforting thought. As it will take over 20,000 years for the message to arrive, the astronomers involved will never know whether they have failed.

Instruments in Space

The era of space astronomy began soon after World War II with
instruments on sounding rockets and large balloons. More than a
thousand satellites or their pieces are now whirling above our
heads, so we will give here only a brief summary. We may
consider that the space age has taken two directions. One is to
reach out physically with space probes to the environs of the
bodies nearest the earth — first the moon and then the distant
planets. The other is to get proper instruments above the earth's
atmosphere to observe the entire universe more fully than can
ever be done from the earth's surface. Both these types of projects
continue to be actively pursued, with phenomenal success. The
first mentioned, being of a more spectacular nature, has had
much more widespread publicity, especially when humans went
aloft. The Surveyor, the Apollo, the Mariner, the Pioneer series,
the Soviet Venera probes landing on Venus, and the Viking to
land on Mars have blazed their trails not merely across the spaces
of the solar system but also across the front pages of newspapers
and covers of periodicals on earth. Their results have been seen
on millions of television screens. When the space probes were
accompanied by astronauts, the pictures have been so spectacular
as to be almost unbelievable, as when we actually saw the Apollo
astronauts walking around on the surface of the moon.

Going less conspicuously about their work is a long procession
of satellites carrying various types of enormously sophisticated
and specialized instruments. (These are in addition to the various
meteorological and commercial satellites.) Among the early
successful ones is the OSO series. The Orbiting Solar Observatory is constructed to study the sun and keep track of the flares and
high-energy particles that can have serious effects on earth. The
remarkable Skylab itself was equipped for solar astronomy
observations. Then there is the series of Orbiting Astronomical
Observatory, OAO. The second, OAO-2, met with great success.
The third, OAO-3 named the Copernicus satellite because it was
launched near the celebrations for the 500th anniversary of his
birth, carried the largest telescope ever put into space. Its 32-inch
reflector specially constructed for ultraviolet work has been

prodigiously successful. The heaviest unmanned satellite (4,900 pounds) ever launched by NASA, it is in a nearly circular orbit with perigee 459 miles above the earth.

Highly specialized projects are already in orbit or in the planning. The satellite launched December 12, 1970, from San Marco launch pad off the coast of Kenya, was given a Swahili name, "Uhuru," which means freedom, in recognition of the kind hospitality of the Kenyan people. It has detected dozens of new X-ray sources in the sky. The HEAO, High Energy Astronomy Observatory, is being designed to continue observation along these lines. The IUE, International Ultraviolet Explorer, is being developed for 1976 or 1977. The IMP series, Interplanetary Monitoring Platform, studies conditions in the depths of the solar system. There is an RAE series, Radio Astronomy Explorer satellite. One was launched into orbit around the moon. Another such satellite will unfold its antennae to a span of hundreds of feet after it gets into space. The Airborne Infrared Observatory will carry a 36-inch telescope for studies in the wavelengths longer than the visible.

The crowning project in the astronomical series of satellites will be the LST, Large Space Telescope. This is planned as a 2-meter reflecting telescope, designed for a minimum life in space of 10 years. The actual life could be many decades. It can be repaired and refurnished while in orbit through visits by the NASA Space Shuttle. It will be remotely controlled with a television-type recording system of much higher sensitivity than photographic film. It can reach objects several magnitudes fainter than it could study from the earth's surface. Furthermore, it can even work with the sun in the sky! This will give 6,000 hours a year observing time, compared with 2,000 for ground-based instruments. The long-exposure images of celestial bodies that can be obtained with this instrument will be at least 10 times finer than from the ground. A resolution of 0.01 seconds of arc can be attained. The best on the ground is 0.5, and that for only a small percentage of the observing time. The LST can also work at wavelengths in the ultraviolet shorter than 0.3 micron, an impossibility for ground-based equipment. Who can predict what unexpected celestial bodies, fine details or cosmic events in space this fantastic tool will reveal!

★

★

★

ELEVEN

The Milky Way

It is easy to understand how a person who hunts for the Milky Way in the environs of one of the world's large cities can reach the conclusion that the whole thing is a figment of the astronomers' imagination. Probably no other celestial object has been pushed more forcibly into obscurity by the blaze of artificial light which has burst forth in our 20th century. It is, in fact, a rather sad commentary on our civilization that, as our technical society grows on earth, it affords most people fewer and fewer opportunities to enjoy nature as it has been until the onset of modern times. One of the few good things about World War II was that the blackouts and the dimouts gave people in the larger cities, such as London, a chance to see the night sky as it had not been seen from that area for more than half a century.

But let us imagine that we have a chance to be far away from city lights, that the moon is out of the sky and the sky is black. Then how does the Milky Way appear? Just as Milton described it, three centuries ago, it is a broad, misty band of light which actually makes a great circle around the sky. Of course the Milky Way cuts off for us at the southern horizon, at the declination where the stars can no longer be seen because of the observer's latitude.

Certainly the early peoples, who could really see the sky, never doubted the existence of the Milky Way. Round and round the world this beautiful apparition was named over and over by different fascinated beholders. Anaxagaros, 550 B.C., and Aratos, whose account of the constellations is the earliest extant, referred to it as "that shining wheel, men call it Milk." The very names galaxy and Milky Way come from these early Greek philosophers. Eratosthenes called it the Circle of the Galaxy, and Hipparchus, who made the first star catalogue, referred to it as the Galaxy.

Not all peoples had the idea of milk, however. To many it was the River of Heaven, or a Great Serpent. Among the Hebrews it was the River of Light. In China and Japan it was the Celestial River or the Silver River. The Romans thought of it as the Heavenly Girdle. Among the Dutch it was the Women's Street, in Germany the Milch Strasse, in Sweden the Winter Street. To the Norsemen it was the Path of the Ghosts going to Valhalla. A similar idea prevailed among the North American Indians, documented by Henry Wadsworth Longfellow in "Hiawatha," when old Nokomis taught little Hiawatha. She

"Showed the broad white road in heaven,
Pathway of the ghosts, the shadows,
Running straight across the heavens . . ."

The Ottawa Indians said it was the muddy water stirred up by a turtle swimming along the bottom of the sky. The Polynesians call it that "Long, Blue, Cloud-eating Shark."

But some of the early Greeks sensed its true character. Democritus, about 460 B.C., and even Pythagoras before him, said it was an assemblage of very distant stars, and Aristotle seems to have agreed. Manilius expressed it thus:

> *"Or is the spatious Bend serenely bright*
> *From little Stars, which there their Beams unite,*
> *And make one solid and continued Light?"*

The proof of its true nature came with Galileo and his first telescopes, which resolved the Milky Way into stars. He wrote in his *Nuncius Sidereus,* that he had "got rid of disputes about the Galaxy — for it is nothing else but a mass of innumerable stars planted together in clusters."

Even though it was the telescope that unveiled the secret of the Milky Way, nevertheless the best over-all view of it is still obtained with the human unaided eye. At one glance your eye can take in a greater sweep and scope of this luminous pathway than you can obtain by studying smaller pieces of it on a larger scale. Perhaps the way to get the maximum enjoyment is to study photographs showing its depths, understand that you are looking at masses of stars piled on masses and then go out under the dark sky and look at its "circling zone . . . powder'd with stars." This is one of Nature's grandest masterpieces. From the tropics and the southern hemisphere, especially at high altitudes, the Milky Way is conspicuously bright. It is actually bright enough to cast a shadow and live up to Pliny's old description of it as "a radiant circle of dazzling brightness."

Not only does the brilliance of city lights prevent many sky-watchers from enjoying the beauty of the Milky Way, but the Milky Way itself plays a little trick that keeps it from being better known. Its position across the sky changes from hour to hour during the night, and from month to month and season to season, because its great circle is inclined 63° to the celestial equator. This is bewildering to the casual observer. We are accustomed to looking for the bright bodies of the solar system, sun, moon and planets, which stay within 28° of the equator. They frequent the limited area of the sky known as the zodiac. They do not move all around the sky as does the Milky Way.

Thus to enjoy the Milky Way fully you need to look in every part of the sky, at different times and different seasons. One of the finest views comes in the early evening in late summer. Then you can trace it from the northeast to the southwest through the overhead regions, from Perseus rising in the northeast, through Cassiopeia and Cepheus. Then when you look overhead at the Milky Way in Cygnus, the Northern Cross, you become really aware of the amount of structure in the Milky Way. You can see a deep black furrow which divides it, leaving it with two branches. At the end of the westerly branch there is an actual gap. The easterly branch then continues on in a real blaze of glory, into the heavy star clouds in Scutum and Scorpius, where it is joined again by the westerly branch in the heaviest star clouds of all in Sagittarius.

Another fine time to trace the part of the Milky Way above northern horizons is in the early evening in late winter. Now it goes from northwest to southeast, reaching high overhead again. Cassiopeia is low in the northwest now, and the Milky Way runs easterly from it through Auriga and Taurus, getting higher in the sky as it passes through Gemini, the Heavenly Twins. As it reaches toward Orion it weakens out, though it leaves symbols of its greatness behind in the form of the Great Nebula in Orion. It is far less conspicuous as it passes through Canis Major in its way to Carina.

In the southern sky, it goes into the parts of the heavens invisible from latitudes north of the tropics. Sir John Herschel, from his studies in South Africa, a century ago gave a beautiful description of the course of the Milky Way through Argus, now called Carina. According to Herschel, it presents a broad, moderately bright, uniform slender stream over the head of Canis Major up to the point where it enters the prow of the ship Argo. Here it again subdivides. The main stream continues southward to 33° south of the equator, where it spreads out broadly and opens into a fanlike expanse almost 20° in breadth in

Carina. The continuity of the Milky Way terminates suddenly. Then it begins again with a similar fan-like group of branches converging on the famous bright star Eta Carinae. Next it courses through a very famous part of the sky — Centaurus and Crux, the Southern Cross. It crosses the hind feet of the Centaur and enters the Cross by a narrow isthmus only 3° or 4° in breadth. This is the narrowest portion of the entire Milky Way. Then it broadens out into a bright mass over the stars Alpha and Beta Crucis and Beta Centauri, and extends almost to Alpha Centauri, the nearest star to us.

One of the most remarkable features of the Milky Way is visible there. In startling contrast to the brilliance of the Milky Way is a dark, pear-shaped vacancy, about 8° in length and 5° across. Even a superficial stargazer can hardly miss it. The early southern navigators gave this startling vacancy the expressive name still used: the Coal Sack. The Coal Sack and the adjacent constellation, the Southern Cross, are two of the objects that are a "must" to see on a trip to the southern hemisphere, or even to the lower latitudes of the northern hemisphere. Photographs fail to do the Southern Cross justice. One of its brightest stars is red, and on a regular blue-light photograph the cross outlines are not accented because this star appears inconspicuous. This region of the Milky Way is one of the most spectacular in its entire course. Near Alpha Centauri it subdivides again, with the main stream passing on into Norma, and thence to the tail of Scorpius, where we can view it in the summer in northern latitudes.

The Coal Sack is the most conspicuous of many dark patches in the Milky Way. From northern latitudes there is no other visible which is more than 1/15 as large as the Coal Sack. In this vast "hole in the sky" only one star is visible to the unaided eye. But here we are not actually looking out into space so deep that there are no more stars. Instead, we are confronted by a dark curtain of dust and gas hanging relatively near us, at a distance of 300 light years, blocking light from the stars behind it.

Though the Coal Sack is larger and blacker than most of the similar regions in the Milky Way, it is typical of a whole group of celestial objects — the dark, diffuse nebulae. These dark patches dotted through the Milky Way contribute to its mottled appearance. These are clouds of dust and gas, made up of common elements like hydrogen and calcium and sodium. They are usually hundreds of light years away from us, and up to a few dozen light years in thickness. A great deal of mass in our galaxy is in the form of such clouds. This may be material left over from the beginning of creation, or it may be material spewed back into space from massive stars in the course of their long lifetimes.

If you look closely around the Milky Way you may see some fuzzy light patches. The outstanding example of these is the central star in the sword which hangs down from the belt of Orion. Even with the unaided eye you can detect that this does not look sharp like an ordinary star. This fuzzy spot is the Great Nebula in Orion. It was the first in a list of six "luminous spots or patches" which Edmund Halley published in 1715. He wrote: "Wonderful are certain luminous Spots or Patches, which discover themselves only by the Telescope, and appear to the naked Eye like small Fixt Stars, but in reality are nothing else but the Light coming from an extraordinary great Space in the Aether . . The first and most considerable is that in the Middle of Orion's Sword, marked with Theta by Bayer in his Uranometria, as a single Star of the third Magnitude; and is so accounted by Ptolemy, Tycho Brahe and Hevelius; but is in reality two very contiguous Stars environed with a very large transparent bright Spot, through which they appear with several others."

The Great Nebula in Orion is a luminous mass of gas, comparable in make-up to the dark nebulae we have been discussing. But this nebula is made to glow by some intensely hot stars embedded in it. The nebula soaks up the ultraviolet light from the hot blue stars in it, and in a sort of fluorescent effect pours light out into space. Until 1927 the light was thought to be coming from an unknown element "nebulium," even though there was no vacancy in the periodic table for such an element. Then Dr. I. S. Bowen, who later became the first Director of the Mount Palomar Observatory, found that the light was being emitted by the ordinary elements of nitrogen and oxygen. Because of the low density and enormous extent of the medium in which they are located, these elements are in a physical state of excitation not achieved on earth. This nebula is at a distance of

1,600 light years from us, with a diameter of 26 light years. It is the brightest object of its kind in our galaxy, in which hundreds of light and dark diffuse nebulae have been catalogued.

Although you can detect the light of the Orion nebula with your unaided eye, powerful telescopes and spectrographs have been needed to reveal its nature. Among the more fascinating results of such studies is the growing conviction in recent years that stars are still being born in this nebula, as well as in other sections of the Milky Way clouds. In such a mass of gas and dust, conditions seem just right for a star to come into being when a globule of matter condenses under gravitational contraction. When the protostar has acquired sufficient energy and heat in its interior, it will then be able to convert hydrogen to helium and provide the energy that keeps stars going. Measures of some of the stars in the region of the Orion nebula indicate that they came into being as stars a million years ago or less. Our earth is at least 5,000 times as old!

As your eye sweeps the broad expanse of the Milky Way, certain areas appear richer than others, like diamond clusters on a gossamer background. In Perseus two such knots especially stand out, set against the rich star clouds there. The most conspicuous group of all, the Pleiades, is on the edge of the Milky Way in Taurus. There are a few groups at some distance from its starry clouds, as in Cancer and Coma Berenices. And in the constellation of Hercules, far from the Milky Way, is another fuzzy star. This is not really just one star but a collection of a quarter of a million stars that form a giant stellar system. These various knots of stars are star clusters, described in Chapter 12.

Just what is the cosmic significance of this wide band of misty light we call the Milky Way? Its significance is that it forms the central plane of our stellar system, our galaxy, which is disk-shaped, with a central bulge. When we look into the Milky Way we are looking into the greatest depth of stars. The over-all diameter of the galaxy is between 80,000 and 100,000 light years. The sun is about 30,000 light years from the distant center in Sagittarius, around which all the stars, star clusters and nebulae are describing orbits. The rotation of the galaxy was first proved observationally by Drs. J. S. Plaskett and J. A. Pearce at the

Dominion Astrophysical Observatory in the 1920's. The sun goes all the way around once in about 250 million years — a cosmic year — with a velocity of 150 miles a second. The sun, dragging all the planets with it, has doubtless been around 15 or 20 times since the earth was formed.

Our galaxy is a spiral. Different objects and groups of stars may be taken as indicating spiral arms. As you search out the Milky Way you may notice a belt of bright blue stars near it but not concentric with it. This is called Gould's Belt, and has a physical significance. These are hot, blue, young stars, at distances of hundreds to a few thousand light years from us. They help show the structure of the galaxy in the sun's neighborhood. Such young stars indicate a spiral arm in Perseus, and another through Orion. Actually, it is easier to trace the spiral arms on galaxies millions of light years away from us than to detect them in our own galaxy when we are inside it!

When you look at the bright stars in the sky you are still somewhat earthbound, and still pursuing a slightly egocentric viewpoint. Most of the bright stars look bright because they are relatively near to us. When you look at the Milky Way, however, you are beginning to get a truer perspective on the universe. The shackles of the sun and the solar system fall away. You are gazing out into an enormous depth of space, and into a system of stars so vast that our sun is but one in 200,000 million. In our galaxy millions of stars must have planetary systems. Some of them may have life comparable to our own. Some of these millions of planets may have extraterrestrial civilizations capable of communicating with us. Though some of the giant ears known as radio telescopes have listened, no sound has yet been received that could be attributed to intelligent life beyond the earth. Only in the present century have men on earth had the knowledge and instruments which could achieve such communication. To accomplish it is one of the greatest challenges facing mankind!

★

★

★
TWELVE

Clusters of Stars

Many a night I saw the Pleiads, rising thro' the mellow shade,
Glitter like a swarm of fire-flies tangled in a silver braid.

TENNYSON, Locksley Hall

Those knots of stars you see in the Milky Way are open clusters. Only a few can be seen with the unaided eye, though a thousand of them have been catalogued in our galaxy. Most of them are inconspicuous. They are embedded in the rich star clouds that make up the length and breadth of the Milky Way. In fact, they seem to form the backbone of the Milky Way.

Open clusters consist of a few dozen to many hundreds of stars. The nearest, the Ursa Major cluster, is about 75 light years away, while the most distant open clusters are many thousands of light years from us. Doubtless others are farther away, at distances too great to detect against the heavy starry background. Some of them are very young, much younger than our earth, less than a million years old. Others with ages of 5 or 6 billion years are older than our solar system. The age of such a cluster may be determined from a spectral analysis of its stars. The stars in a cluster were born together out of the primordial cloud. Together they have been traveling through space ever since, with speeds considered high by human standards.

Most outstanding of all open clusters is the Pleiades, one of the most photographed objects in the realm of the stars. The Pleiades look so much like a little dipper that someone learning the constellations for the first time may mistake them for the constellation of Ursa Minor, the Little Dipper, which is really more difficult to see. The Pleiades are located near the conspicuous V in Taurus, rising in the east in the early evening in the fall. As the months march on, this cluster moves into the western sky after sunset in the spring.

For at least 4,000 years, mankind has admired and even worshiped this little group of stars. They are woven into the folklore and legends of many peoples, and are among the first stars to be mentioned in literature, in Chinese annals of 2357 B.C. They were recorded in the Hindu lunar zodiac before 1730 B.C. And Job asked: "Canst thou bind the sweet influence of Pleiades, or loose the bands of Orion?" The Pleiades were considered so important that many of the early peoples began their year with November, the Pleiad month. The risings and culminations on the meridians of the Pleiades set the dates for feasts. The festivals of All Hallows' Eve and All Saints Day commemorated them. In South Africa they are referred to as "hoeing stars." Their last visible rising after sunset is celebrated with rejoicing as signifying the "waking-up time" for farmers.

The actual names of the cluster and the individual brightest stars come from Greek mythology, in which the Pleiades were the seven daughters of Atlas and Pleione. Alcyone is now the brightest of the seven sisters in the cluster and known as "the lucida." At the time the Chinese first recorded Alcyone, it was near the position of the vernal equinox, but precession has now carried it 24° north. In some accounts Maia, the first born and most beautiful of the sisters, is referred to as the brightest. Have the names been interchanged, or has the relative brightness of the stars changed? Asterope, Taygeta, Celaeno, Electra and Merope complete the group.

Over and over again the number "seven" is used in their description. In China they were called the Seven Sisters of Industry. Sancho Panza called them the Seven Little Nanny Goats. A popular term though the centuries was Man with Six Chickens. Among various peoples of the world there is a persistent legend of a lost Pleiad. Only six stars — not seven — are now relatively conspicuous to the unaided eye. It is quite probable that one Pleiad has faded in brilliance in the last two millennia, but which one is not known. Electra is surrounded by a legend that her light dimmed when she witnessed the

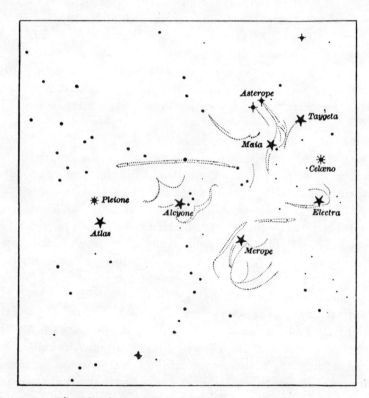

Figure 34. *The Pleiades as portrayed by Charles A. Young.*

destruction of Ilium. And in recent decades Pleione has undergone some remarkable episodes, casting off shells of gas. The bright blue, hot Pleiades stars are outstanding for the high speeds at which they are whirling on their axes, with velocities as much as 200 miles a second at their equators.

Persons with keen vision have seen as many as eleven Pleiads with the naked eye. This is the number on a chart made by Maestling, Kepler's teacher, before the invention of the telescope. Actually, the Pleiades cluster may have as many as 500 stars. It is over 30 light years in diameter, at a distance of about 400 light years. The stars have been traveling together for 150 million years in their paths around the center of our galaxy.

The Pleiades stars are embedded in beauteous nebulosity, which is invisible to the unaided eye. The nebula around Merope can be seen in a telescope, but photographs are best for revealing these delicate features. The nebulosity shines by the reflected light of the Pleiades stars. This was proved by Dr. E. C. Slipher in 1912. For 21 hours on three consecutive nights he put the light of the nebulosity through a spectrograph on the 24-inch telescope at the Lowell Observatory, Flagstaff, Arizona. The nebula is like a painted sign illuminated by electric light, and not like a luminous neon sign. Despite its diffused beauty, the nebula is a very poor reflector, reflecting about 1 percent as much light as would a sheet of white paper. The Soviet astronomer Dr. V.A. Ambartsumian has shown that the nebulosity in which many open clusters are embedded is really a part of the general galactic structure. It may not particularly belong to the stars associated with it.

Not far from the Pleiades in the sky, and in the same constellation of Taurus, is the group of stars known as the Hyades. They form part of the V of this constellation, which is conspicuous throughout the fall, winter and early spring. Centuries ago this group was called the "rainy Hyades," perhaps because it rises in the autumn when the fall rains come.

The Hyades is one of the nearest clusters to us, at a distance of about 120 light years. There are probably 350 stars in this cluster, grouped in a sphere 40 light years in diameter. About half of this estimated number have been scientifically identified as cluster members.

The Hyades is called a "moving cluster" because the motions of its individual stars are large enough so that they define a spot in the heavens known as the convergent point. The cluster appears to be streaking toward it at 27 miles a second. This point is near the position of the bright red supergiant Betelgeuse in Orion. But do not expect to see this motion with your unaided eye! To observe even the largest stellar motions on the sky is comparable, as Dr. A.D. Thackeray of South Africa has said, to watching the growth of a fingernail at the distance of one mile.

Astronomers rate the Hyades as one of the most important clusters in the sky. It has been studied so thoroughly by many different astronomers, with all the powerful instruments at their disposal, that its distance is precisely determined. Its stars now can be taken as standards, in terms of temperatures, luminosities, ages and various other characteristics. These standards can then be applied to appraise other stars of similar types and measure their distances throughout the universe.

Another cluster of stars, fainter than the Hyades, may be seen in the constellation of Cancer the Crab. Cancer is the constellation for which the Tropic of Cancer is named, because in it the sun used to reach the northernmost point of its journey. This group is Praesepe, the Beehive or Manger, which appears to the unaided eye as a faint patch of light. Centuries ago the Chinese called it "the exhalation of piled-up corpses" — not one of the lovelier star names, certainly. This "whirling cloud," as the Greek authors called it, was first resolved by Galileo into a multitude of more than 40 small stars. About 200 cluster stars are now identified. The brightest stars are 6th magnitude, just visible to the unaided eye under good conditions. Praesepe is at a distance of 500 light years from us.

A curious feature is that Praesepe is similar in all its properties to the Hyades, which is far from it in the sky and much nearer to earth. Its velocity through space is equal to that of the Hyades. It is going 27 miles a second, and in the same direction, to the eastern edge of Orion. It seems probable that the clusters had a common origin in the distant past, about 1,000 million years ago.

Another moving cluster is that in Ursa Major. In contrast to faint Praesepe, most observers are familiar with the individual

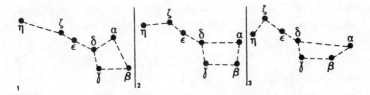

Figure 35. *Not all the stars of the Big Dipper are physically held together in a cluster, so it will come apart in time. At left, 50,000 years ago; at right, 50,000 years hence.*

stars of the Ursa Major Cluster. This cluster, at a distance of only 75 light years, is so near to us and has so few stars that it does not look like a cluster. It consists of a nucleus of 11 stars within a sphere of 30 light years, including five central stars in the Big Dipper. Measurements show a common motion for these stars. Surrounding the nucleus is an Ursa Major stream, with stars at distances up to 400 light years from it. The sun is in the volume of space occupied by this larger Ursa Major cluster so that its members, one of which is the bright star Sirius, appear scattered all over the sky. The sun is not considered a member of the cluster, however.

Although it has few members, it will hold together as a cluster for several thousand million years. Its stars have such a high velocity, 11 miles a second, that they are little affected by any others they pass. The Big Dipper itself, however, will fall apart in some tens of thousands of years, as shown in the diagram, because not all of its stars belong to the cluster and share its motion. Some will stray in time.

Another faint patch in the constellation of Coma Berenices is a sparse and feeble open cluster. This constellation, Berenice's Hair, is visible in the spring and summer. It is the only constellation named for a real, historical figure. There *was* a Berenice and her hair became of historical importance. She was the wife of Ptolemy Euergetes of Egypt, in the 3rd century B.C. When her husband went off to war, as an evidence of her devotion to him, she cut off her tresses and laid them on the altar for his safe return. Unfortunately, someone created consternation in the court by stealing them. Happily, the court soothsayer proved equal to the occasion. He overcame the horrible experience by pointing to this little patch of light in the heavens and saying, "There! There is the hair of Berenice in the sky, immortal forever."

The little group of stars that form the Coma Berenices cluster is readily visible to the unaided eye. While the constellation was named in early times, only in the present century was it first realized that the stars formed a physical cluster. There are only 37 members scattered over a relatively large area, 7° in diameter. These stars are now shown to be traveling together, heading in the same general direction as our sun, but going only two-thirds as fast, at eight miles a second. The cluster is 250 light years away. A feeble little cluster like Coma will not hold together forever. Probably it has already run a fair share of its course, and is headed for disintegration as a cluster. This little system has had its day and will cease to be.

To avoid confusion, it should be mentioned that the term "Coma Cluster" has a second meaning. This is an enormous cluster of galaxies 30 million light years from us, concentrated within the boundaries of this constellation. Dr. F. Zwicky of Mount Palomar Observatory has published counts of galaxies in the Coma Cluster from the 48-inch Schmidt telescope, National Geographic — Palomar sky survey. These indicate that about 10,000 galaxies make up this cluster. On these plates many of the galaxies have a tiny image, like a slightly fuzzy star. Great care is taken in these counts, says Zwicky, because "to mistake a particle of dust on the photographic plate for a whole galaxy of stars is certainly going from the sublime to the ridiculous."

Now it happens that the north pole of our galactic system also lies in Coma Berenices, which means that the stars in our galaxy are definitely thinning out in this direction. So a long exposure photograph of this region of the sky can show more galaxies per square degree than individual stars in our galaxy! In this direction we are coming to the end of the stars in our galaxy, but the realm of the other galaxies becomes increasingly prominent.

In the Milky Way in the constellation of Perseus is a rich, bright cluster, which you can see as a double knot. Perseus is technically visible all nights of the year for latitudes north of 40°

N. because it is a circumpolar constellation, but in the early summer evenings it is too close to the northern horizon to be readily seen. This cluster in Perseus is well named the "Double Cluster." The two individual clusters that make it up were earlier designated as h and Chi Persei. (Between Chi and Eta Persei lies the radiant of the Perseid meteors, mentioned in Chapter 2.)

The first record of this cluster comes from Hipparchus, 150 B.C., who called it "a cloudy spot." At a distance of about 7,000 light years, this double cluster is one of the younger clusters in the sky, only a few million years old, far younger than our earth. It contains very hot blue stars and some red supergiants. Some of these stars are 10,000 times as luminous as our sun. Dr. P. Th. Oosterhoff of the Leiden Observatory, Holland, has catalogued over 2,000 stars in the region of these clusters, but fewer than half of these are actual cluster members. They are exceedingly rich clusters, but since they merge into the rich star cloud behind them, it is difficult to separate the clusters from the clouds.

In the sprawling constellation of Hercules is another misty patch visible to the unaided eye. Hercules is above the horizon at night from late spring till middle autumn. Edmund Halley first drew attention to this patch, along with the Great Nebula in Orion and four others. Halley published his list of six nebulous objects in the Proceedings of the Royal Society of London, with an interpretation of their significance.

"The Sixth and last was accidentally hit upon by M. Edm. Halley in the Constellation of Hercules, in the Year 1714. It is nearly in a Right Line with ζ and η of Bayer, somewhat nearer to ζ than η. This is but a little Patch, but it shews it self to the naked Eye, when the Sky is clear, and the Moon absent.

"There are undoubtedly more of these, which have not yet come to our Knowledge, and some perhaps bigger, but though all these Spots are in Appearance but small, . . . yet since they are among the Fixt Stars, . . . they cannot fail to occupy Spaces immensely great, and perhaps not less than our whole Solar System."

This "little Patch" is a cluster so rich in stars that it is a very different type of cluster from the open. This one is a globular star cluster, a glorious system of a quarter of a million stars at a distance of around 20,000 light years. The designation Messier 13 comes from that catalogue of 107 hazy objects which the French astronomer, Charles Messier, aided by his colleague Pierre Méchain, made half a century after Halley. Messier 13 is often referred to as "the great globular cluster in Hercules." In this constellation there are two other globular star clusters, neither visible to the naked eye.

Messier 13 certainly does occupy a space far larger than our whole solar system! It takes light about 300 years to go from one edge to the other of this great cluster, compared with less than a day to traverse the planetary domain of our solar system. Inside such a cluster the stars are much closer together than are stars in the neighborhood of our sun. An observer at the center might see a thousand stars as bright as Sirius scattered around his sky.

In addition to Messier 13, three other globular clusters may be seen with the unaided eye. The first to be discovered is one in the constellation of Sagittarius, found by Abraham Ihle in 1665 and later known as Messier 22. With declination −24° it is low in the southern sky from northern middle latitudes, and is better seen the farther south you go. This is a very massive cluster. One estimate is that it contains 2 million suns. Messier 22 is the northern vertex of an almost equilateral triangle, which it forms with Lambda and Phi Sagittarii. Phi is the most westerly star of the northern quadrilateral that makes up part of the constellation and Lambda is the star that seems to connect the two quadrilaterals. Messier 22 is on the edge of the Milky Way, only 8° from its central plane.

The two brightest globular clusters are located even deeper in the southern sky. These are Omega Centauri and 47 Tucanae, the two most powerful clusters in our galaxy. The star in Centaurus that had been lettered as Omega by Bayer was found to be a nebulous patch by Halley in 1677 when he journeyed south to the island of St. Helena to catalogue the southern stars. With a declination of −47° it is visible as a 4th-magnitude object, south of latitude about 40° N. — the farther south the better. It is located between Xi and Gamma Centauri, a little nearer to Xi and slightly north of the line joining them. This cluster has an estimated 2 million stars in a sphere of diameter 600 light years.

The fuzzy star in the constellation of Toucan assigned Flamsteed number 47 is a magnificent system of an estimated 1.3 million suns. In the sky it is seen near the Small Magellanic Cloud and frequently appears alongside it on photographs. It is west of the northern edge of the Small Cloud, and is near a line from Beta Hydri to Beta Toucanae, and one-third of the way. With a distance of 20,000 light years it is only one-ninth of the way to the Cloud, discussed in the next chapter.

Dotted around our galaxy are about 130 globular clusters, most of them faint and distant. The nearest is about 7,000 light years from us, the most distant 400,000. They outline the extent of our galaxy because they are the most distant objects we can recognize that belong to it. They are not concentrated to the plane of the Milky Way as are the open clusters. Rather, they are dotted about the entire sky. However, one-third of the total number known are concentrated in and around the heavy star clouds in the Sagittarius-Scorpius region. This concentration of mass led Dr. Harlow Shapley, at Mount Wilson Observatory early this century, to propose that the center of the galaxy lies there. Subsequent observations of many kinds have indeed confirmed that the galactic center is behind these rich star clouds.

Successful investigations of globular clusters require large telescopes and present many challenging facets. Many of these clusters have variable stars — because of their great distance none visible to the unaided eye. Many of the variables are Cepheids, one kind of yardstick that can give the distance to the cluster. A study of the relative brightnesses and colors of the hundreds or in a few cases the thousands of the brightest stars yields a color-magnitude diagram. From such a diagram much valuable information can be derived, including the age of the cluster. Globular clusters prove to be as old as our galaxy itself — about 14 billion years by current reckoning. A study of their spectra shows them to be deficient in elements of atomic number higher than the first two elements, hydrogen and helium. They were formed in the very early stages of our galaxy, before these elements existed. Some stars had to be in existence before these elements of higher atomic number could be built up in their fiercely hot interiors, especially in those stars which burst forth as supernovae.

Theoretically, their beautiful symmetry is due to the rotation of the individual stars about the center of the cluster, their common center of gravity. Recently this has been proved for the cluster Omega Centauri by British astronomers of the Royal Greenwich Observatory using the 74-inch telescope of the Radcliffe Observatory in South Africa.

Not merely are the individual stars of a globular cluster circling its center, but the clusters themselves are swinging in great orbits around the center of the galaxy. These orbits take them hundreds of millions of years to complete. Some of the most distant clusters may be at the extreme end of their orbits — apogalacticon. Our sun also is circling the distant center, once around about every 250 million years, a cosmic year. Our earth is never in the same position in space again. It is possible that vast eons from now the sun and earth may be much closer to some globular cluster. Someday on our earth a race of men may have a sight that we can never have. They may come so close to one of these great globes of stars that the brilliance of the star-studded night sky will be dazzling.

★

★

★

THIRTEEN

The Limits
of the Universe

For tho the Giant Ages heave the hill and break the shore
And evermore make and break and work their will —
Though world on world in myriad myriads roll round us
Each with other powers, and different forms of life than ours
What know we greater than the soul?

TENNYSON, "Ode on the Death of
the Duke of Wellington"

Our galaxy is huge, but even more enormous are the distances to
the galaxies beyond. Nevertheless, it is a remarkable fact that the
human unaided eye has the power to penetrate these vast
distances and see some of the other galaxies. Three galaxies far
beyond our own can be seen easily with the unaided eye, the
Large and Small Magellanic Clouds and the Great Galaxy in
Andromeda.

The Magellanic Clouds

Dwellers in the southern hemisphere are fortunate because they can see the two galaxies nearest us. These are the Large Magellanic Cloud and the Small Magellanic Cloud, which are located deep in the southern celestial hemisphere. They are in a region of the sky with exotic constellation names: the Phoenix, Flamingo, Flying Fish and others. The Large Cloud is mostly in the constellation of Dorado the Goldfish, and the Small Cloud is in Toucan. To the unaided eye the diameter of the Large Cloud is less than 8°, and of the Small, less than 4°. Together with the celestial pole they form the vertices of an equilateral triangle. They are located so near the South Pole, both within 25° of it, that south of the tropics they can be seen every night of the year. This gives travelers who make only brief trips to the southern hemisphere a chance to enjoy them too.

The Clouds were well described by Alexander von Humboldt, the last scientist to attempt a total description of the universe. "The two Magellanic Clouds arrest the attention of the traveler, as I have myself experienced, in the most forcible manner by their brightness, their remarkable isolated position and their revolution at unequal distances around the southern pole. The two Magellanic clouds, of which the larger covers forty-two and the smaller ten degrees of the celestial vault, produce at first sight, as seen by the naked eye, the same impression as would be made by two detached bright portions of the Milky Way of corresponding dimensions. In strong moonlight the smaller cloud disappears entirely, while the larger one only loses a considerable portion of its light. The Magellanic Clouds are neither connected with each other nor with the Milky Way by any perceptible nebulous appearance. The smaller nubecula is situated in what, excepting the vicinity of the star-cluster in Toucani, is a kind of starless desert, the larger Magellanic Cloud in a less scantily furnished part of the celestial vault."

Despite the name assigned to them, the Magellanic Clouds were known centuries before the time of Magellan. As early as the middle of the 10th century A.D. the Arabians had a name for the Large Cloud, the White Ox, el-bakar. The celebrated

astronomer Dervish Abdurrahman Al-Sufi of Tai, a town in the
Persian Irak, says: "Below the feet of Suhel [Canopus] there is a
white patch, which is not seen either in Irak [in the region of
Bagdad] nor in Nedschd [the northern and more mountainous
part of Arabia], but is seen in southern Tehama between Mecca
and the point of Yemen, along the shore of the Red Sea." The
position of the white patch relative to Canopus agrees with that
of the Larger Magellanic Cloud. In the time of Abdurrahman Al-
Sufi, according to von Humboldt who computed the precession,
the Large Cloud was visible as far north as 17°N. latitude. The
two clouds, he declares, might have been seen throughout the
whole of the southwest of Arabia. The extreme southern point, at
Aden on the Straits of Ba-el-Mandeb, is in 12° 45'N. latitude.

The Clouds were certainly known to other navigators before
the voyages of Magellan because they were mentioned in the year
1515 both by the Florentine Andrea Corsali in his voyage to
Cochin and by Petrus Marthy de Anghiera, Secretary to
Ferdinand of Aragon. Six years later Pegafetta, who accompanied Magellan, first mentioned the "nebiette" in the journal of
the voyage, January, 1521, when the ship *Victoria* made her way
from the Patagonian Strait into the South Pacific Ocean. Early
navigators, first the Portuguese and later the Dutch and Danish,
called these "Cape Clouds" perhaps because of Table Mountain
at the Cape. A small cloud at its summit may portend a storm, a
phenomenon dreaded by seamen.

John Herschel, son of William, made the first systematic
investigation of the Clouds, studying the different types of
objects they contain. In 1825 in the southern hemisphere near
the Cape, with a "twenty feet reflector" he began the
continuation of his father's "Sweeps of the Heavens." The
explicit title of his first work on them, published in 1833, is *First
approximation toward a catalogue of objects in the Magellanic Clouds,
containing 919 stars, nebulae and clusters in the greater and of 244 in
the lesser nubecula intended as preparatory to the construction of charts
of the nubeculae and to the future execution of drawings of them in
detail.*

Von Humboldt was quick to appreciate Herschel's observations of stars, star clusters and nebulae in the Clouds, and

attempted to sense their cosmic significance. "From the complex character of the nubeculae therefore, they ought not to be regarded (as is too often done) as extraordinarily large nebulae, or as detached portions of the Milky Way." Von Humboldt also drew attention to a now famous object in the nubecula major, "a nebula noticed as early as by Lacaille (30 Doradus, No. 2941 of Sir John Herschel) which is without a parallel in any part of the heavens."

It is surprising that the early observers looking through telescopes could acquire so much factual knowledge of the distant Clouds. But with the advent of celestial photography their structure has become much more thoroughly understood. The first photographs of the Clouds were taken in 1890 near Sydney, Australia, by Mr. H. C. Russell, F.R.S., with a portrait lens of 6 inches aperture. On a high hill well outside the city he exposed his plates all night, for seven or eight hours on various nights. From the photographs thus acquired he concluded that both the Large and the Small Cloud showed spiral structure and hence were complex spiral nebulae.

Since those days the number of photographs taken of the Clouds runs into the hundreds of thousands. Even in the early years of the 20th century, however, astronomers were still puzzled as to the real nature of the Clouds. The fact is they are so relatively near to us that their similarity to the spiral nebulae — as the galaxies beyond our own were first called — was effectively masked. It is an excellent example of the adage that you cannot see the woods for the trees.

Nowadays the study of the Clouds is being pursued on a wide variety of fronts, and we are beginning to understand our nearest neighbors among the galaxies. The Large Cloud is 160,000 light years away, receding at 165 miles a second. The Small Cloud is 190,000 light years away, receding at 100 miles a second. Different sorts of distance determinations give slightly different values. The Large Cloud is a flattened system, with an observable off-center bar or axis across it, as certain types of galaxies have. Its diameter is about 30,000 light years and its mass 6,000 million solar masses. Its rotation has been observed, the outer parts rotating in a period of 250 million years. The Small Cloud, with a diameter of 25,000 light years containing 1,500 million solar masses, seems to be seen almost edge on. Its period of rotation is probably similar to that of the Large.

Old stars make up about 90 percent of the mass of both Clouds, according to Dr. Bart Bok of the University of Arizona. These are the stars known as Population II. The young stars (Population I), prodigal of energy are most conspicuous on the Cloud photographs, but are really just the frosting on the cake. Thousands of variable stars have been found in the Clouds. In fact, it was from a study of the variables in the Small Cloud that Cepheids came into importance. When Miss Henrietta Leavitt of the Harvard Observatory studied some variables in the Small Cloud in 1912, she found that the brighter ones had the longer periods. Since the Clouds are so far away, the relative brightness of stars we see in them is in the same ratio as their actual brightness. Shortly thereafter, Shapley applied this period-luminosity relation to a study of the size of our galaxy and the distances of the globular clusters.

In addition to stars, as Herschel found, the Clouds also contain a wealth of fascinating objects — open and globular clusters, loose groupings of very bright stars termed constellations, gaseous nebulae and planetary nebulae. A few novae have appeared.

That large nebula, 30 Doradus, which so fascinated von Humboldt, continues to intrigue 20th-century observers too. Also known as the Loop or Tarantula nebula, with its central association of very luminous young blue stars, it is the dominant complex of the Large Cloud. A map made with a radio telescope set at the fundamental wavelength of the neutral hydrogen atom, 21 cm, shows this huge nebula is composed mostly of hydrogen. If it were located as near to us and in the position of the Great Nebula in Orion, it would subtend an angular diameter of 25° to 30° in the sky and fill the constellation of Orion and cast shadows on earth! Eons past, the Clouds may have been nearer us. Did prehistoric living things then have a chance to bask in the light of the Tarantula nebula?

At the distance of the Clouds any star less than 50 million years old will be observed at present within half a degree of its place of origin. The Magellanic Clouds are hence an almost ideal laboratory for the study of star birth and evolution.

It is true, as von Humboldt noted, that there is no visible connection between the Clouds. But radio astronomy gives another picture. Observations in 1954 at a frequency of 40 kilocycles per second by F. J. Kerr of Australia and colleagues showed that the two Magellanic Clouds are actually embedded in a common envelope of neutral atomic hydrogen. But no connecting link with our galaxy appears, though there is a similarity of chemical abundances of elements in the Clouds and in the galaxy.

The mystery of the origin and birth of the Clouds is still that for us — a mystery. Eventually, observations will pile up to an extent that we can turn the universe backward in time and make informed guesses as to the far distant past.

Great Galaxy in Andromeda

While from the mid-northern latitudes we are deprived of the sight of the two nearest galaxies, we do have the privilege of looking at the most distant object that the eye can see in the universe. The Great Galaxy in Andromeda, Messier 31, is about 2.2 million light years away. For decades it was called (and sometimes still is) the Great Nebula in Andromeda. However the term "nebula" is more properly reserved for gaseous masses. The main body of the Andromeda galaxy is composed of thousands of millions of stars. It has gaseous masses in it, but these are only a small part of it.

To see this in the sky you need a dark, moonless night away from city lights. First, locate the Great Square of Pegasus. Then start from the star in the northeast corner of the Square. As noted in Chapter 8, this star is actually Alpha Andromedae. From this star as your eye goes north and east you see a line with the next two brightest stars, Beta and Gamma Andromedae. When you reach Beta (Mirach), head up north and a little west from it. About 4° in this direction is Mu Andromedae, a 4th-magnitude star, and another 4° northwest brings you to the Andromeda galaxy. If sky conditions don't permit you to see Mu Andromedae, it is unlikely that you can see Messier 31. Of course it will appear only as a misty, faint spot, as did the Great Globular Cluster in Hercules, but with a larger angular diameter. But what an object it is, at the end of your span of 2 million light years.

Sky-watchers have seen this little patch for at least 1,000 years. Just as the Persian astronomer Abdurrahman Al-Sufi noticed the Large Magellanic Cloud, so did he see this patch of light in the girdle of Andromeda near the star Gamma. Later on, Spanish and Dutch constellation charts of the 14th and 15th centuries gave dots at its position. But little attention was paid to it until after the invention of the telescope, when Simon Marius (known as Mayer of Genzenhausen) studied it.

Marius relates that, on December 15, 1612, he discovered a fixed star different in appearance from anything he had seen before. It was near the third and northernmost star in the girdle of Andromeda. With the unassisted eye it looked only like a small cloud. Through the telescope he could find in it nothing resembling stars. All that could be distinguished was a white shining appearance, brighter in the center and fainter toward the margin. The whole was about a fourth of a degree in breadth, and resembled a light seen from a distance shining through semi-transparent horn (as in a lantern).

This description applies today. However, under exceptionally favorable conditions of sky and altitude, the nebula can be a striking object. In the 1890's W. H. Pickering while at the Harvard Station at Arequipa, Peru, at 8,000 feet above sea level, described it as conspicuously bright and larger than the full moon.

For more than two centuries the nebula was regarded as "starless." As telescopes improved markedly in the mid- and late 19th century some astronomers became convinced that they were resolving the outer edges of the nebula into stars. Just as for the Magellanic Clouds, photography, with the build-up of the image over long exposures, helped to show the true structure. Dr. Isaac Roberts at Maghull, near Liverpool, had a silver-on-glass reflector of 20 inches aperture, and took a three-hour exposure of the Andromeda nebula on October 1, 1888. A convoluted structure was apparent, along with two attendant nebulae, which appeared like satellites around it.

Decades of astronomical observations of the Great Andromeda Galaxy have revealed its nature and characteristics. This galaxy is larger and more massive than our own. On the sky it covers an area 1½° by 4½°. At its distance this corresponds to a diameter of 180,000 light years. Actually it is nearly circular, but appears oval to us because of its projection. Apparently we are looking at it almost edge on. Its spiral arms show its plane to be slightly warped, possibly by tidal interactions with two of its companions.

The Andromeda Galaxy is composed of very much the same type of objects that make up our own galaxy. There are star clusters (266 of them catalogued) and young associations (188 of them). There are gaseous emission nebulae (688 of them), mostly in a ring of neutral hydrogen gas with a radius of about 30,000 light years, which surrounds the central bulge of Messier 31. Great star clouds are there, dotted with knots of bluish stars, which are the counterpart of our open clusters. They also contain hundreds of variable stars, mostly of the important Cepheid type. Many ordinary novae have been found. Apparently, they appear there at the rate of 30 per year.

The total mass of the Andromeda Galaxy is estimated to be 300,000 million solar masses. This is determined from the application of Kepler's laws to the measured velocity of rotation of the system. For just as our galaxy is rotating about its massive center, so is Messier 31. At a distance of about 45,000 light years from the center, parts of the great spiral are rotating with a velocity of 190 miles a second. The heart of the nucleus, with a radius of about 20 light years, is in very rapid rotation almost like a solid body. It is spinning around in a period of a million years, while the outer parts take several hundred times as long to go around. In general, the more distant the parts of the galaxy are from the center, the smaller is their velocity in their orbit, just as for the planets in our solar system.

In 1885 the Andromeda galaxy was the scene of a great event, which took decades to be understood. On August 17 of that year M. Ludovic Gully of Rouen noticed a new bright star of the 9th magnitude, near the center, only 16 seconds of arc southwest of it. The night before, Tempel and Max Wolf, keen observers, had seen nothing unusual about the nebula. By September 1 the star had reached 7th magnitude, not far below naked-eye visibility. Then it declined rapidly until it reached the limit of visibility of those days, the 16th magnitude, in the 26-inch refractor at Washington. Although Dr. A. A. Common of England had taken a photograph of the Andromeda galaxy a year before the star appeared, no photograph was taken when the star was bright, and no photographic record of it remains. The spectroscope, beginning to come into astronomical use then, revealed curious and rather contradictory features of this new object, S Andromedae.

For years there was controversy as to whether a star of such great apparent brightness could be an actual member of the distant nebula where the other stars were thousands of times fainter. If the new star were an actual member, its intrinsic luminosity would have to be so great as to be almost unbelievable. Yet the probabilities were very low indeed that it could be a foreground star. Eventually the new star was accepted for what it really was: the first of the mighty class of supernovae to be recognized.

Those two attendant systems, which Roberts noticed on his early photographs, are elliptical galaxies, smaller companions of the great Andromeda galaxy. Neither is visible to the unaided eye. Both had been seen and recorded many years before Roberts photographed them. The brighter one had been catalogued by Messier as 32, just following his object 31. The other one, first noted by William Herschel and his sister Caroline, is known by its number in Sir John L. Dreyer's first catalogue of nebulous objects and star clusters, the New General Catalogue, NGC 205. In 1974 three more companions, very faint, were discovered by Dr. Sidney van den Bergh, bringing the total in the Messier 31 system to six galaxies.

Our galaxy is not alone in space. The two Magellanic Clouds, the Great Galaxy in Andromeda and its five companions are seven of the companions of our own galaxy. Our galaxy is one of a group of at least 20 galaxies which, in the violent motions of space, are remaining together. These galaxies are known as "The Local Group." (Astronomers are a trifle blasé about distances when they apply the term "local" to a group of objects strung out along 2 million light years of distance from end to end.) The Great Galaxy in Andromeda is the most magnificent companion that our galaxy has. It is the most massive with the greatest intrinsic luminosity and the largest diameter. Our galaxy comes second, but a relatively close second. More than half the group have diameters of less than 10,000 light years. They are dwarf elliptical or irregular galaxies on a scale much smaller than the Magellanic Clouds. Besides the Magellanic Clouds, the Andromeda galaxy and our own, only one has a diameter greater than 10,000 light years. That is the great spiral galaxy in Triangulum. Its diameter is 50,000 light years and although its distance is only slightly greater than the distance of the Andromeda galaxy, 2,400,000 light years, it cannot be seen by the naked eye. The reason is partly that it is not so intrinsically luminous and partly that it is a much more open spiral, with less concentration toward a bright nucleus.

On a proportionate scale, the galaxies in our Local Group are closer together than are the stars in the sun's neighborhood. That is, the distance separating our galaxy from the nearest external galaxy, the Large Magellanic Cloud, is about twice the diameter of our galaxy. But the distance from our sun to the nearest star, Proxima Centauri, 4 1/3 light years, is about 30 million times the diameter of the sun.

The Galaxies Beyond

Beyond our tight and cozy Local Group are millions of other galaxies of all types. On photographic plates their images range in size from large spirals just beyond our Local Group, with distances of a few million light years, to small, fuzzy images that are hard to differentiate from single stars, at distances as great as 10,000 million light years. As was described for the Coma Cluster, more of them can be seen in the directions around the galactic poles where our own galaxy has less material in it. These regions offer the best windows to the universe beyond.

The dividing line between the galaxies in our Local Group and those outside comes mainly from a study of the velocities of these objects. The measurement of these velocities has been one of the greatest observational challenges of the last half-century in astronomy. In the early years of this century measures by V. M. Slipher and later by Edwin Hubble showed that some galaxies were approaching us and some going away, at velocities up to several hundred miles a second. These measures come from the positions of the lines in the spectrum that shift as the object moves toward or away from us. This is an application of the famous Doppler principle in sound, whereby the pitch of a moving source (train bell or whistle) actually changes as the source moves toward or away from the listener.

As the Doppler shift was studied for fainter and fainter galaxies, the remarkable discovery was made that all of these were running away from us — the velocities were all shifted to the red, indicating recession. The more distant the galaxy, the greater its velocity of recession. Longer and longer exposure times were required to obtain velocity determinations as far out in space as possible. Eventually, Milton Humason of the Mount Wilson Observatory exposed the 100-inch telescope on the same galaxy all night several consecutive nights. The galaxy was so far away that it could not be seen in the telescope, and Humason had to trust to dead reckoning for its position. The greatest red shift yet measured for a galaxy was by Dr. Hyron Spinrad in 1975 with the Lick 120-inch telescope. The radio galaxy 3C-123, at a distance of 8 billion light years and five to 10 times the size of our galaxy, has a Doppler shift 45 percent that of the speed of light.

The greatest measured red shifts now belong to quasi-stellar objects, quasars, those mysterious objects first identified in radio telescopes because of their enormous energy in the radio region. Unlike galaxies, these cannot be resolved into stars. At first they seemed almost unbelievable. Even now there is still scientific discussion as to whether the enormous shift of the quasars' light to the red is due to a velocity shift. So far, this is the best explanation.

So the picture of a quasar is of a body with the mass of millions of suns, confined to a space a few light years in diameter (the distance from the sun to the nearest star!) and moving away from us with enormous measured velocities. The greatest so far is OQ172 in Boötes with a red shift caused by a velocity amounting to 91 percent that of light. Perhaps quasars are one stage in the evolution of a galaxy — a gigantic nuclear reaction at the start.

Apparently the limits of the explorable universe will be set for us when we reach objects that are traveling away from us with the speed of light. Theoretically, at least, such objects are forever invisible, and the universe will cut off for us at the distance where these objects are. Could their light reach us by coming around the universe from the other direction? Though scientists have tested this idea, so far there is no observational evidence that this does happen.

Age and Origin of the Universe

When and how did the universe start? Different methods of dating the universe agree that it is within 12 to 18 billion years old. Studies of the evolution of stars in the oldest objects in our galaxy, the globular star clusters, give these values. And analysis of the chemical distribution of atoms in the universe shows that heavy elements began to originate in that era of the distant past. Critical to the age determination are the velocities of expansion of galaxies from their site of formation, and these yield the highest value.

All well-founded theories of the origin of the universe rest on what is called the cosmological principle. In essence this states that any observer should see the universe the same in all directions, and the universe should look the same to all observers wherever they are.

The Steady State theory of the British scientists H. Bondi, T. Gold and F. Hoyle demands the continuous creation of matter. This is necessary to keep the density of the universe constant when the red shifts of galaxies show that the universe is expanding. The amount of matter to be created seems small — unless one remembers the sizes and time intervals involved. It amounts to one atom a century in a cube of space which has between 1,000 and 2,000 feet to a side. While there is now abundant observational evidence for the creation of stars from matter, there is none as yet for the creation of matter itself.

The Big Bang model of the universe, favored by physicist George Gamow, starts with a primordial fireball of incredibly high density (Ylem) at some time in the past. Expansion goes on to infinity, perhaps at an accelerating rate. Many of the chemical elements were formed in only the first half-hour after the Big Bang. A remnant background field temperature of 3° above absolute zero was predicted from the cooled fireball. A. A. Penzias and R. W. Wilson of the Bell Telephone Laboratories actually observed such background radiation at a temperature of 2.8° above absolute zero just as R. H. Dicke and associates of Princeton University were predicting its existence.

The Pulsating Universe theory begins in a similar way, but the expansion reaches a limiting radius and then falls back on itself.

The Big Bang theory seems now to be most in accord with observational results published in 1974-75. Dr. Alan Sandage and his colleagues of the Hale Observatories have concluded a program on the distant galaxies outlined by Dr. Edwin F. Hubble in 1951. This program has pushed the great 200-inch telescope to its limits. The rate of recession of the galaxies is determined to be 55 ± 5 kilometers per second per million parsecs of distance from us. (A parsec is 3¼ light years.) From these values an age of the universe is deduced as 18 billion years.

The observations lead Sandage to the conclusion that the expansion will never stop — the galaxies will continue to speed outward in all directions. A similar conclusion is reached by California and Texas scientists studying the amount of matter in the universe. They find there is no more than a tenth of the amount of mass which is necessary to pull the expansion to a halt, and, to use the current expression "close the universe." The determinations of mass have been made in several different ways which corroborate one another.

Scientists try to be very objective in their work, but these conclusions seem to go against the philosophical grain of the astronomers who have produced them. A fading and escaping collection of galaxies is not as pleasing an idea as a pulsating or steady state universe, which goes on forever. The idea that "the Universe has happened only once" leaves a great deal to be desired.

Does this vast and magnificent universe in which our earth is but a speck make man seem small and insignificant? Not at all. Just think of it. Such a tiny part of us as the human unaided eye can see to a depth of 2 million light years. Even more miraculous, humans have devised instruments to probe the universe to its physical limits. And crowning triumph, the human mind can comprehend the objects which our eyes and instruments reveal.

APPENDIX I

Amateur Organizations, Planetariums and Observatories

Headquarters of Amateur Organizations

Royal Astronomical Society of Canada, National Office: 252 College St., Toronto, Ont. M5T 1R7. Centres: Calgary, Edmonton, Halifax, Hamilton, Kingston, London, Montreal, Centre d'Astronomie (Montreal), Niagara, Ottawa, Quebec, St. John's, Saskatoon, Toronto, Vancouver, Victoria, Windsor, Winnipeg.

American Association of Variable Star Observers (AAVSO), 187 Concord Ave., Cambridge, Mass. 02138

Association of Lunar and Planetary Observers, J. Russell Smith, Secretary, Waco, Texas 76710

Astronomical League, Science Service Building, 1719 N St. NW, Washington D.C. 20036

British Astronomical Association, Burlington House, Picadilly, London W1 VONL

International Union of Amateur Astronomers, Secretary, 93 Currie St., Hamilton, Ont.

Société Astronomique de France, 28, rue Saint-Dominique 75007 Paris, France

Planetariums in Canada

Calgary Centennial Planetarium, Calgary, Alta.
H.R. MacMillan Planetarium, Vancouver, B.C.
McMaster University Planetarium, Hamilton, Ont.
Seneca College Planetarium, Willowdale, Ont.
Laurentian University Planetarium, Sudbury, Ont.
Dow Planetarium, Montreal, P.Q.
Queen Elizabeth Planetarium, Edmonton, Alta.
Manitoba Museum of Man and Nature, Winnipeg, Man.
McLaughlin Planetarium, Royal Ontario Museum, Toronto, Ont.
Coast Guard Academy, Sydney, N.S.
School of Fisheries, St. John's, Nfld.

Major Observatories in Canada

Optical
David Dunlap Observatory, Richmond Hill, Ont. University of Toronto. 74-inch reflector and other telescopes
Dominion Astrophysical Observatory, Little Saanich Mountain, Victoria, B.C. National Research Council of Canada. 72-inch reflector and other instruments
University of Western Ontario. 48-inch reflector at Elginfield, Ont., and other instruments in London, Ont.
About ten other universities have observatories of moderate size.

Radio
Under the auspices of National Research Council of Canada
Algonquin Radio Observatory, Lake Traverse, Algonquin Park, Ont.
Dominion Radio Astrophysical Observatory, near Penticton, B.C.

APPENDIX II

Celestial Log

Foreword to Log

No satisfaction in astronomy is greater than seeing some of the beautiful objects, or some of the exciting happenings, with your own two eyes. Here is a list of some of the sights available. For a few of them luck, as well as diligence, will be necessary. Many of them are very easy, especially when you are away from big city lights. It is a sad commentary that modern civilization with its bright lights and indoor attractions has detracted from one of man's greatest birthrights — the ability to enjoy the sights in the eternal heavens. Cavemen were better off than we are in this respect!

After you have tried some of these observations you may find yourself so intrigued by them that you want to continue observation as a research project. For example, variable stars quickly capture an observer's interest. For such research endeavors, a telescope or, at the least, binoculars are necessary for most observations. If you wish to continue in this way, then contact the headquarters of the appropriate organization listed in Appendix I for information and instructions.

Object	Observing Location	Date	Time	Remarks

ATMOSPHERE:

  Twilight bow
  Green flash
  Halo around sun
  Halo around moon
  Corona around sun
  Corona around moon
  Mirage

EVENTS IN ATMOSPHERE:

  Aurora
  Sporadic meteor
  Annual shower
  Meteorite in museum
  Meteorite crater on
    earth

MOON:

  Harvest moon
  Hunter's moon
  High meridian
    altitude full moon
  Low meridian
    altitude full moon
  Earthshine on the
    moon
  Moon's motion
    among stars,
      2 observations,
      4 hours apart
  Occultation of star
    by moon
  Occultation of
    planet by moon
  Eclipse of moon
      Partial
      Total

SUN:

  Sunrise or sunset
    point near
      Vernal equinox
      Summer solstice
      Autumnal equinox

Object	Observing Location	Date	Time	Remarks
Winter solstice				
Naked eye spot				
Eclipse				
Partial				
Total				
Corona				
Prominences				
Baily's beads				
Shadow bands				
Animal behavior				

PLANETS:

Object	Observing Location	Date	Time	Remarks
Mercury, greatest elongation E. greatest elongation W.				
Venus, greatest elongation E. greatest elongation W. greatest brilliancy Transit, 2004 A.D.				
Mars, opposition				
Jupiter, opposition				
Saturn, opposition, rings open opposition, rings closed				
Uranus, opposition				
Vesta, opposition				
Planet's loop of retrogression 3 observations, days or weeks apart				

STARS:

Object	Observing Location	Date	Time	Remarks
Motion of circumpolar constellations 2 observations, 3 hours apart				
Time by the stars				

Object	Observing Location	Date	Time	Remarks

PERFORMING STARS:
    Algol at minimum
    Mira at maximum

MILKY WAY:
    Cygnus rift
    Sagittarius clouds

STAR CLUSTERS:
    Open star clusters
        Pleiades
        Other
    Globular star
      clusters:
        Messier 13
        Other

OTHER GALAXIES:
    Messier 31,
      Andromeda galaxy
    Large Magellanic
      Cloud
    Small Magellanic
      Cloud

SIGHTS THAT REQUIRE LUCK AS WELL AS PERSISTENCE:
    Blue sun (or green)
    Blue moon (or
      green)
    "Fire alarm" sunset
    Spectacular meteor
      shower
    Fireball
    Detonating fireball
    Brilliant comet
    A new star

USE EXTRA PAPER FOR THESE DRAWINGS:
    (1) Sunrise or sunset points
    (2) Moon's motion among
        stars
    (3) Planet's loop of
        retrogression

APPENDIX III

Time by the Stars

There are several methods by which you can tell time by the stars. One that is easy and popular is by the use of the Pointers of the Big Dipper, which we have met already. The Big Dipper can be used as a sky clock on any clear night.

The procedure is as follows. Consider the northern sky as a giant clock face with Polaris, the Pole Star, as the center, and the line joining it with the Pointers as the hour hand. (See diagram.) This sky clock has no minute hand. On this clock face estimate the time to the nearest quarter of an hour. On the diagram the time would be about 9:45.

Now we will derive Eastern Standard Time, but the time can be derived for any zone by applying the necessary correction to Eastern Standard. To go from the sky clock to Eastern Standard Time you need to apply a simple rule of thumb. To your estimated time, add the number of months and fractions of a month that have elapsed since the New Year. For May 13 we will call this 4½. Multiply this sum by 2, and subtract it from 52¼. This magic number in the formula is easy to remember because it happens to be the number of weeks in a year. The

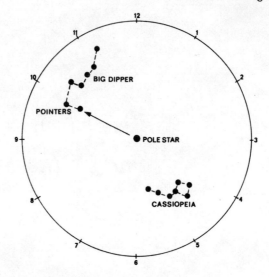

Figure 36. *One example of a star clock.*

resulting figure is the approximate Standard Time for a place on the standard meridian.

For example, on May 13 with the above appearance of the star clock,

Estimated time	9¾
Months since Jan. 1	4½
Sum	14¼
Multiply by 2	28½
Subtract from 52¼	23¾

The Eastern Standard Time is approximately 23¾ hours, or 11:45 P.M., and Daylight Time is 12:45 A.M. With a little practice it is easy to estimate the star clock time to a quarter of an hour.

APPENDIX IV

Bright Stars and Constellations

If you know the star pictures and memorize the following rhyme, you will ever have at hand for reference on clear nights a reliable timepiece, compass and calendar.

The numbers above the star names indicate consecutively the months of the year in which these respective objects rise about the first of each month in the eastern sky. In addition to 1st magnitude stars the rhyme refers to the head of Capricornus, the Sea-Goat, the Great Square of Pegasus, and Orion's Belt. All except Arcturus rise between 9 and 9:30 P.M. Arcturus rises at 10 P.M. February 1.

> First Regulus^1 gleams on the view,
> Arcturus,^{2} Spica,^{3} Vega^4 blue,
> Antares,^{5} and Altair,^{6}
> The Goat's^7 Head, Square,^{8} and Fomalhaut,^{9}
> Aldebaran,^{10} the Belt^{11} a-glow,
> Then Sirius^{12} most fair.*

Eight months of the year are identified by the position of the Big Dipper at 9 P.M. In April and May it is north of the zenith. During July and August it is west of north. In October and November it lies close to the northern horizon and in January and February it is east of north with the "Pointer Stars" highest.

*From *Field Book of the Skies,* by William Tyler Olcott. Revised by R. Newton Mayall and Margaret W. Mayall (New York: G.P. Putnam's Sons, 4th ed., 1954), p. 460. Reprinted by permission of the publisher.

Supplementary Reading List

Almanacs

The Observer's Handbook. Royal Astronomical Society of Canada, 252 College St., Toronto, Ont. M5T 1R7.
Old Farmer's Almanac. Yankee, Inc., Dublin, N.H.

Atlases and Maps

MENZEL, DONALD H. *A Field Guide to the Stars and Planets.* Boston: Houghton Mifflin Co., 1964.

NORTON, ARTHUR P. *Norton's Star Atlas.* Cambridge, Mass.: Sky Publishing Corporation, 16th ed., 1973.

OLCOTT, WILLIAM TYLER. Revised by R. Newton Mayall and Margaret W. Mayall. Field Book of the Skies. New York: G.P. Putnam's Sons, 4th ed., 1954.

UNIVERSITY OF TORONTO. *Sky Maps.* David Dunlap Observatory, Richmond Hill, Ont. L4C 4Y6. Set of six, $1.00 postpaid.

Periodicals

Astronomy. 640 N. LaSalle St., 6th Floor, Chicago, Ill. 60610.
The Griffith Observer. Griffith Observatory, 2800 East Observatory Rd., Los Angeles, Calif. 90027.

Journal of the Royal Astronomical Society of Canada. Royal Astronomical Society of Canada, 252 College St., Toronto, Ont. M5T 1R7.

Mercury, the Journal of the Astronomical Society of the Pacific, 75 Southgate Ave., Daly City, Calif. 94015.

Science News. Science Service Inc., 1719 N St., N.W. Washington, D.C. 20036.

Scientific American. Scientific American, Inc., 415 Madison Ave., New York, N.Y. 10017.

Sky and Telescope. Sky Publishing Corporation, 49-50-51 Bay State Rd., Cambridge, Mass. 02138.

Books

BAKER, ROBERT H., and LAWRENCE W. FREDRICK. *Astronomy.* New York: Van Nostrand Reinhold Company, 9th ed., 1971.

HAWKINS, GERALD S., with JOHN B. WHITE. *Stonehenge Decoded.* Garden City, N.Y.: Doubleday & Company, Inc., 1965.

HAWKINS, GERALD S. *Beyond Stonehenge.* New York: Harper and Row. Toronto, Fitzhenry and Whiteside Limited, 1973.

MAYALL, R. NEWTON, MARGARET W. MAYALL, and JEROME WYCKOFF. *The Sky Observer's Guide: A Handbook for Amateur Astronomers.* New York: Golden Press, 1971.

MAYALL, R. NEWTON, and MARGARET W. MAYALL. *Skyshooting.* New York: Dover Publications, Inc., 1968.

— *Sundials, How to Know, Use, and Make them.* Cambridge, Mass.: Sky Publishing Corporation, 2nd ed., 1973.

On UFO's

HYNEK, J. A. *The UFO Experience.* Chicago: Henry Regnery Co., 1972.

MENZEL, D. H. *Flying Saucers.* Cambridge, Mass.: Harvard University Press, 1953.

MENZEL, D. H., and L. G. BOYD. *The World of Flying Saucers.* Garden City, N.Y.: Doubleday & Company, Inc., 1963.

Index

$1:

The Stars Belong to Everyone

Helen Sawyer Hogg

Illustrated with
Diagrams and Photographs

You don't need a telescope to enjoy a moonrise, admire the Evening Star, locate the Big Dipper, watch an eclipse or follow the arch of the Milky Way. But this book will enable you to more fully appreciate these and the many other fascinating objects and events in the skies overhead.

You may be surprised at the variety of subjects Dr. Hogg covers — everything from an account of the Great Moon Hoax to the conundrum for the maximum number of Sundays possible in February. Here, you will find answers to many of the questions often asked about astronomy — questions like:

- How long will the sun last?
- When will Halley's comet return?
- Is there really such a thing as a blue moon?
- What is the most distant object you can see without a telescope?
- What is a pulsar? A black hole? A supernova?
- What does "expanding universe" mean?

In addition to the celestial sights you can see with just your own two eyes, you will also learn about the world's most powerful telescopes and what they have revealed about deep space. Thus, with this book, you can travel the universe from neighboring Mars to the far-distant Andromeda Galaxy, for indeed, the stars belong to everyone.